Beyond COINCIDENCE

Beyond COINCIDENCE

*Stories of Amazing Coincidences and
the Mystery and Mathematics
That Lie Behind Them*

MARTIN PLIMMER
and BRIAN KING

Thomas Dunne Books ♒ St. Martin's Press
New York

THOMAS DUNNE BOOKS.
An imprint of St. Martin's Press.

www.stmartins.com

Book design by Jennifer Ann Daddio

Library of Congress Cataloging-in-Publication Data

Plimmer, Martin.
   Beyond coincidence : stories of amazing coincidence and the
mystery and mathematics behind them / Martin Plimmer &
Brian King.
      p.   cm.
   ISBN 0-312-34036-2
   EAN 978-0-312-34036-0
   1. Curiosities and wonders.   2. Coincidence—Miscellanea.
   I. Title.

AG243 .P58   2005
031.02—dc22

                                              2005048492

First published in the United Kingdom by Icon Books

First U.S. Edition: January 2006

10   9   8   7   6   5   4   3   2   1

To the memory of

P E T E R   R O D F O R D

who, coincidentally,

taught us both

"Any coincidence," said Miss Marple to herself, *"is always worth noting. You can throw it away later if it is only a coincidence."*

—AGATHA CHRISTIE

CONTENTS

One:

Coincidence Under the Microscope

Two:
Coincidence on the Rampage

One

COINCIDENCE
UNDER
THE
MICROSCOPE

THE COSMIC YES!

I know we've met before.

Let's talk. I'm sure we'll find things in common—broad strokes first: language, race, nationality, gender, front door color, love of Italian food, the desire to jump in puddles . . . this book!

Soon we'll narrow the focus: places we have both visited, times we have nearly met, events at which we were separated only by people we didn't know, people we know in common. . . .

The more we look the more we'll find. Before long we'll discover we have lived in the same town, or attended the same school, were born in the same hospital, had the same accountant, the identical dream. . . . Perhaps we traveled on a bus together—perhaps we have rubbed shoulders on a bus!

The idea makes us shiver. Why? Does it mean anything? Not objectively. After all, we rub shoulders with strangers on buses every week and it doesn't make us shiver.

But now we are no longer strangers. And now, knowing each

other, we can see that we have always known each other. If, after this meeting, we become friends, we will think those former meetings and almost meetings very significant indeed. We will call them coincidences, but we will think them more than that. The trick of being personal is in the nature of coincidence. It is always particular, always subjective, always to do with us. Fate has singled us out. You and me. It's that specialness that makes us shiver.

We all do this, especially if we want to like each other. We frisk each other for links. We're like synchronized swimmers in search of a routine. We relish connections, and we're a highly connected species. If it were possible to map all human activity, drawing lines between friends and relatives, departures and arrivals, messages sent and received, desires and objects, you would soon have a planet-size tangle of lines, growing ever denser, with trillions of intersections.

Each intersection is an association waiting to be noticed as a coincidence, either for its own sake or when yet another intersecting line passes through it. Coincidence is commonplace. It's everywhere. But we are only aware of those intersections that are meaningful to us. Paul Kammerer, an Austrian biologist of the early nineteenth century, said that these are manifestations of a much larger cosmic unity, a force as powerful as gravity, but that acts selectively, bringing things together by affinity. We only notice its peaks, which are like the ripples on the surface of a pond.

Just what this force affecting us all might be, we don't know. Suggestions include a higher universal intelligence, gods and aliens (both mischievous and benign), a psychomagnetic field, the controlling power of our own thoughts, or a universal system of parallel universes operating in different dimensions from ours. That's easy to say, hard to understand, and impossible to prove.

Back on the ground, all that most of us are aware of when we notice a coincidence is that it provokes one of those shivers. It might be just a tremor of the imagination, but the more unlikely the coincidence, the more the sensation of invisible fingers running down the back.

Bolt-from-the-blue coincidences involving events or material objects—like bumping into a long-lost friend in a foreign town, or finding a toy in a yard sale that you once owned as a child—can move even the most skeptical in ways that are difficult to define.

What is it about coincidences that grabs at the emotions? It's the frisson of being touched by something outside of yourself. It's a sense of being chosen. One minute you are stumbling through quotidian chaos, trying to find a phone that is ringing or maneuvre a stroller up the stairs of a bus, the next you are in a lacuna of clarity, where every disparate thing—events, objects, your own thought processes—appears to be bent to the same end. For a second the suspicion that you are tiny and insignificant, and the Universe arbitrary and terrifying, disappears. You are part of a great cosmic YES!!!

Research has been done suggesting that people who notice coincidences most tend to be more confident and at ease with life. Every coincidence they experience—even the minor ones—confirms their optimism. They know that things will happen to them, that somewhere in a second-hand Tijuana bookshop there is likely to be the one remaining signed copy of their father's only novel, that in an apartment in Hong Kong, waiting to be found, very likely lives the sister they have not been told about, that the signet ring they lost in Chicago is sitting right now on the bottom of Lake Michigan awaiting the astonished angler's casual hook. They are routinely alert to coincidence, certain that at any moment, in the raffle of infinite possibilities, their lucky number will be called. For these people the world really is a smaller place.

Let's think about that book in Tijuana for a second. Let's say it was written by my father. If you came across it as you were browsing books, you wouldn't think anything of it; after all, rare bookshops are full of rare books—it's in their nature. But if I were to find that book, open it and recognize the signature of my late father, carelessly scrawled there when he was a younger man than me, with the world and all its manifold possibilities lying at his feet, the experience would be loaded with poignant significance. Just what it signified exactly, I would be hard pressed to explain, nevertheless my world would look very different from what it had looked like a few moments before. I might even have to sit down.

Laurens van der Post says in his book, *Jung and the Story of Our Time* (about Carl Jung, the Swiss psychologist who defined the concept of synchronicity), "Coincidences, instinctively, have never been idle for me, but as meaningful, I was to find, as they were to Jung. I had always had a hunch that coincidences were a manifestation of a law of life of which we are inadequately aware ... [Coincidences], in terms of our short life, are unfortunately incapable of total definition, and yet, however partial the meaning we can extract from them, we ignore them at our peril."

We know that coincidences are events, objects, and thoughts slung together by the wind—a matter of pure chance—but because they resonate so intimately we think them more than that. Who's to say, when the coincidence has a profound effect on our lives, that we are wrong? Look at what happened to the English Woman *Margaret Muir*.

During the Second World War when she was living in Cairo, Margaret Muir befriended a serviceman who was stationed there. It was one of those friendships that so easily could have become more, but both of them were married, so they decided that was that. After the war they only saw each other every two years, over lunch at the Guard's Club, to catch up on each other's lives. They were still attracted to each other, but now they both had families.

Fourteen years went by and lately the meetings had grown even less frequent. One day Margaret experienced a strong urge to ring her friend. She hadn't seen him for a long time, but decided it probably wasn't wise. So she occupied herself with the crossword in the paper instead. The desire to telephone her friend didn't leave her though.

One of the first clues she solved was "Where a tree and a metal will meet." The answer was ASHORE. This made her pause: her friend's name was Ash. She reached for the telephone, but once again she resisted the urge.

A subsequent clue was an anagram of the word ashore: "Vocally sick, anagram of six across." As she hit on the answer—HOARSE—the desire to telephone her friend became irresistible. She picked up the phone and dialed his office. His secretary answered. "I'm sorry but Mr. Ash died two years ago."

Margaret was shattered by the news, but needing to know more she dialled the number of a mutual acquaintance. "Yes," they said. "We should have told you. He had cancer of the throat. Before he died he got hoarser and hoarser."

Margaret found the experience astonishing, as well she might. She's a rational woman and she still doesn't know how to explain it. "It continues to baffle me in my seventies," she says.

A skeptic would deny the significance of Margaret's crossword experience. He would say it was pure chance that the clues that day related to something in her life. He would say that she had selected, out of thousands of possible elements affecting her life at the time, the two that echoed an idea that was clearly absorbing her. What about the other crossword clues—what did they have to say on the matter of her friend? What if she had decided to do the bridge puzzle instead? Would that have spoken to her as well?

The skeptic may be right, but he can't deny the power of the experience. And anyway, he's missing the point. Margaret's crossword coincidence may not have been meant, in the sense that someone or something had orchestrated it, but it was certainly meaningful. It resulted in her finding out about her friend's death for one, and it made her feel connected to him, despite the separating barriers of time and distance.

Here's another curious fact: the coincidence is meaningful to the reader as well. The reason we respond positively to accounts of coincidence, ever eager to give them the benefit of the doubt, is because they make such good stories. They have the resonance of myth and fairy tale, with their dramatic shifts of fortune, their spectacular life-and-death events and their many enchanted objects and preoccupations: crossword clues, rings, keys, addresses, numbers, and dates. There is a ritualistic quality to them because of their complex storylines: the histories, character traits, dream and thought processes that have to be established and explained in the correct sequence before the coincidence climax can be appreciated. A good coincidence story has the gravitas of Greek drama, the difference being that it is true.

We, the authors of this book, have been acutely aware of this quality while writing it. So we have not skimped on stories. There are almost two hundred in part 2 and others scattered through the book. Some are old classics that wouldn't let us leave them out (and which, like myths, thrive on the retelling); many are told here for the first time.

Carl Jung called coincidences "acts of creation in time." The sheer potency of the stories, and the emotional catharses and transformations they wrought in some of their subjects, testify to that.

All writers have a working arrangement with coincidence. Few novelists are too proud to insert a dramatic coincidence in order to tart up a lackluster plot. Without coincidence comedy routines would be deadly serious. Allegory and metaphor work by linking together two normally unconnected ideas in order to startle the reader into seeing something they thought they knew in a different light. When the poet Stephen Spender describes electricity pylons crossing a valley as "bare like nude, giant girls that have no secrets" he is utilizing the visual energy of something entirely unrelated to pylons in order to shock the reader into a sense of blatant and gauche vulgarity. Strictly speaking metaphors aren't coincidences, as they are manmade, but they work the same trick: fusing unrelated entities to power a revelation.

An interviewer once asked the writer Isaac Bashevis Singer how he could possibly work in such an untidy study. The interviewer had never seen such a cramped and confusing place. Every ledge in the room was taken up with teetering stacks of paper and books piled up on top of each other. It was perfect, said Singer. Whenever he needed inspiration, a pile of papers would fall off a shelf and something would float to the floor that would give him an idea.

There is one kind of coincidence for which all writers have a healthy respect. It manifests itself when you are researching a subject and relevant facts appear everywhere you look. Carl Jung called it the Library Angel and grateful writers leave offerings to it by their bookshelves at night.

When Martin Plimmer was researching this book, he started looking for information about neutrinos—particles so small scientists have never seen them. He was interested in ways we might connect, not just with each other, but also with the physical world and the universe beyond. Neutrinos, which originate in stars and shower the Earth constantly, seemed to suggest a medium for universal intimacy, as they pass right through us, and then through the Earth beneath us, without stopping for the lights, flitting through the empty space in atoms as though there were nothing there. Martin had never heard of neutrinos, but pretty soon the air was thick with them.

He opened a newspaper at random and there was a story about neutrino research. A novel he was reading offered an interesting neutrino theory. When he turned on the television, there was former president Bill Clinton talking about them in a speech. When he looked down at his fingertip, a billion of them were passing through it every second. Funny how he'd never noticed them before. It was as though the whole world had gone neutrino flavor.

Again, you could say that all this stuff is always out there waiting to be noticed, part of the barrage of information that passes before our overloaded senses every day. Our attention is selective; we see only what preoccupies us at the time. That week neutrinos were big, so neutrinos were everywhere. The world appeared to be bent to Martin's current obsession, and the search engine of the gods was suggesting ideas, links, and information.

None of this theory about the barrage effect of information applied to novelist and historian Dame Rebecca West when she was searching for a single entry in the transcripts of the Nuremberg trials. She had gone to the library and was horrified to discover hundreds of volumes of material. Worse, they were not indexed in a way that enabled her to look up her item. After hours leafing through them in vain, she explained her desperation to a passing librarian.

"I can't find it," she said. "There's no clue." In exasperation she pulled a volume down from the shelf. "It could be in any of these." She opened the book and there was the passage she needed.

It would be nice to think that something less fickle than chance was at work here. Did the item want to be found? Did her mind, with the right concentration of energy, *read* which volume it was in? Or did the Library Angel lend her a helping hand? Whatever helped her, it distributes its blessings impartially: for Rebecca West, a session transcript out of thousands from a trial of mass murderers; for a devout Muslim fisherman in Zanzibar eager for evidence of God's greatness, a fish with the old Arabic words "There is no God but Allah" discernible in the patterning on its tail.

If coincidences cluster around preoccupations, imagine how they fall over to please you when the subject is coincidence itself. This book's origins lie in a five-part series, *Beyond Coincidence*, made by Testbed Productions for BBC Radio 4. No sooner had we started our research than so many coincidences happened around us that we began to feel like we were being stalked. Newspapers fell open at accounts of coincidences, potential contributors, at the moment we rang them, were interrupted in the act of writing about coincidence (or so they told us, but then they were in the coincidence business, too).

On the way back one day after interviewing a woman who was doing psychic research, we pulled off the road for lunch. Suddenly Martin began talking about the design of a car he had seen. This was unusual because, uniquely among men, Martin is not interested in cars. Normally he can't differentiate one model from another; he can barely remember the make of his own. This time he'd been so struck by the car he'd made a mental note of the make. It was an Audi.

"And if you could afford to buy one of those cars," said Brian, "would you like to own one?"

"Well, yes," said Martin, considering a novel idea. "I think I would."

"Then I know just the place you should go," said Brian, and he pointed through the window behind Martin's back. Across the road was a car showroom called Martin's Audis.

Early on in the research process Martin read Arthur Koestler's famous 1972 book, *The Roots of Coincidence*. First he poured himself a hot bath. It happened that at that time he was sorting out old vinyl records that he hadn't played for years. He'd been systematically playing these records to see if they were worth keeping or not. At the point when he settled into the bath with Mr. Koestler, Mickey Jupp, a little-known artist from England, started singing in the background.

There was Martin, out of his depth in scientific experiments into paranormal phenomena, specifically to test whether there is such a thing as telepathy. Early research was carried out in Russia by a scientist called Bechterer, who, afraid the authorities might consider his work too frivolous, disguised the telepathy part by calling it "biological radio." No sooner had Martin read that phrase than in the other room Mickey Jupp sang the words "nature's radio." It was a song about telepathy between lovers: "You won't have to tell me, because I'll already know / I'll have heard the news on Nature's Radio."

Now, if Martin were an Ancient Greek he would have regarded this as a good omen. Actually he did anyway. It feels good when two pieces of like phenomena shake hands in your bathroom, particularly when they have judged your mood so well.

Later Martin and Brian were conducting random street interviews. The sixth person stopped turned out to have devoted his life to the celebration and recording of coincidence. By this time it did not seem odd at all. Coincidences? We could call them up at will! Perhaps what we should have been doing was concentrating our powers on winning the lottery.

So far so benign. We've tended to think of coincidences as having good intentions, though as neither random events nor the actions of gods (depending on your standpoint) are necessarily friendly, there's no reason why they should be so restricted. Of course, unhappy coincidences happen all the time.

If a woman were to urinate in your suitcase because it resembled her unfaithful husband's (you snigger, but it has happened), you would feel as though you were trapped in a real-life Larry David routine.

If all the passengers on a 747 jumbo jet happened to pack a small anvil in their hand luggage, the effect wouldn't be beneficial to any of them and they probably wouldn't be able to tell the story either, though the rest of us would be fascinated to read about it in a newspaper, or a novel, or watch it take place in a film. The pattern of coincidence engages us, even when it involves hardship or tragedy, and even victims of bad coincidence may experience a compensating sense of being included or chosen. Sometimes it is better to be noticed, even if we suffer by it, than to be ignored.

We have included lucky stories and unlucky stories in this book, funny stories, sad stories, violent stories, and romantic stories. We love coincidence so much that a misguided assumption prevails, that if a story contains one then it is, by definition, interesting. Many coincidence books have been written with that lazy principle in mind. It may be true, but apart from revisiting a few classic stories such as the *Titanic* prediction and the Lincoln and Kennedy similarities (shiver) we've worked hard to find stories that are interesting in their own right. We threw away a lot of stories that went: "I traveled to Thailand for my summer holiday and I met a woman who used to go to school with my brother in Philadelphia." On the other hand if we've left the odd one in, it's because the woman wore purple lipstick or knew how to wolf whistle.

Enjoy the book. You should, as we share an interest in coincidences. Maybe we'll run into each other again one day.

But then again, realistically speaking, maybe we won't, because, despite all the shivering, we have to be realistic, otherwise we'd sit at home all day waiting for lucky gold bars to fall out of badly loaded planes onto our lawn. It's all very well being one with the great cosmic YES! but everyday life has to go on. As comedian Steven Wright says: "It's a small world, but you wouldn't want to paint it."

. 2 .

WHY WE LOVE COINCIDENCE

Mrs. Willard Lovell locked herself out of her house in Berkeley, California. She had spent ten minutes trying to find a way back in when the postman arrived with a letter for her. In the letter was a key to her front door. It had been sent by her brother, Watson Wyman, who had stayed with her recently and taken the spare key home with him to Seattle, Washington.

Most of us have locked ourselves out of our houses at some time or other. Many of us will have received a key through the post in a letter. Very few of us will have had them happen to us with such exquisite timing. How would you have felt if that had happened to you? Pretty special probably. There you are, trying to cope with an unwelcome and vexing predicament when suddenly, like magic, the solution is handed to you on a plate, or in this case, in a letter. The denouement is so neat and perfectly resolved it makes the anguish of the beginning worthwhile. And what a story to tell your friends!

All the world loves coincidences. We are attracted to their pattern and order—their symmetry. We can even become addicted to them, seeking them out in the most unlikely places. The more unlikely a coincidence, the more we savor it.

And the more remarkable the coincidence, the more the sense that it must have some sort of meaning. Coincidences suggest some sort of controlling, godlike hand is at work—smoothing out the chaos in our complicated lives.

For many people these very personal experiences of synchronicity, or meaningful coincidence, can border on the religious. A major survey conducted in the United States in 1990 asked people to describe spiritual or religious-like experiences they had had. A large majority cited "extraordinary coincidences."

Stephen Hladkyj spent several years studying coincidences experienced by fellow students at the University of Manitoba. He found that first-year students at the university who scored "high on a measure of synchronicity," or were alert to synchronicity or meaningful coincidences in their lives, also scored higher on a self-rated measure of psychological health and had generally adapted well to their first year of college life.

He concluded that people who are alert to coincidence in their lives—particularly personal coincidence—tend to see the universe as a friendly, orderly, responsive place, and consequently develop a general sense of well-being. Coincidence, it seems, is good for us.

It gives us a delicious frisson of pleasure to know that a balloon released by ten-year-old Laura Buxton in her garden landed 140 miles away in the garden of another Laura Buxton aged ten. When coincidences like these occur it is as if people and places, times and events have been choreographed in a way that defies the law of probability.

If we were distant observers from Mars, the story would have no significance whatsoever. A little girl releases a balloon. Some time later it comes down in a garden somewhere else and is picked up by another little girl. Nothing exceptional here. Children are attracted to balloons after all, and balloons do go up and come down again. But seen from an Earth-bound point of view, and particularly from the perspective of the two Lauras, it takes on an entirely different meaning. It sends a shiver down the spine. Because it's personal, you see. It's *so* personal.

The fact that the principals in this story are little girls adds poignancy, but the coincidence would have been just as extraordinary had they been old men, or millionairesses, or even Martians. Coincidence makes no distinction between class, religion, or creed. It happens to us all, whoever we are, whatever we believe. We all are subject to its weblike embrace. To the axiom that only two things are certain in life, death and taxes, must be added a third—coincidence.

Even after death, coincidence can strike.

> *Charles Francis Coghlan, one of the greatest Shakespearean actors of his time, was born on Prince Edward Island on the east coast of Canada in 1841.*
>
> *Coghlan died suddenly on November 27, 1899, after a short illness while performing in the port town of Galveston, Texas, in the southwest of the United States. The distance was too great to send the body back home, so it was interred in a lead-lined coffin in a granite vault in a local cemetery.*
>
> *On September 8, 1900, a great hurricane struck Galveston—hurling huge waves against the cemetery and shattering vaults. Coghlan's coffin was washed out to sea.*
>
> *It floated into the Gulf of Mexico, then drifted along the Florida coastline and out into the Atlantic where the Gulf Stream took over and carried it north.*
>
> *In October 1908, fishermen on Prince Edward Island saw a long, weather-beaten box floating ashore. After nine years and three thousand five hundred miles, Charles Coghlan's body had come home. His fellow islanders reburied him in the graveyard of the church where he had been baptized.*

Coincidences of the kind that befell Charles Coghlan, or the lucky key lady, Mrs. Lovell, are immensely attractive to us. They appeal to our innate need for order and pattern. They make us seem less small and insignificant and the universe less terrifying and aimless. Even the most hard-bitten skeptic can find comfort in the most modest of coincidences. Our preference, naturally enough, tends to be for benign coincidences—particularly when we are the recipient of the good fortune. But malign coincidences are also interesting to us—as long as they are viewed from a distance:

> *Jabez Spicer, of Leyden, Massachusetts, was killed by two bullets in an attack on an arsenal on January 25, 1787, during Shays' Rebellion. He was wearing the coat his brother Daniel wore when he, too, was killed by two bullets on March 5, 1784.*
>
> *The bullets that killed Jabez Spicer passed through the holes made by the bullets that had killed his brother Daniel three years earlier.*

When coincidence does dump misfortune on our doorstep, we at least have the compensation of feeling that we have been singled out by fate for special attention. Most commonly, however, coincidences are modest, unthreatening, and cheering. When we take our dog for a walk in the park and meet a fellow dog walker with an identical dog—with the same name—it brightens our day a little.

How often have you been thinking about someone when the phone rings, and it is that person? Does it not create a frisson of pleasure, a warm feeling? When such things happen we often conclude that we are blessed with the gift of extrasensory perception or are party to some sort of psychic connectedness. We don't like to think it is simply the laws of chance and probability at work. We see such events as transcending physical laws, as being beyond coincidence, beyond the normal—paranormal, in fact. A more rational explanation would be too dull, too meaningless.

It is much more interesting to believe coincidences, particularly the more unlikely events, are predetermined in some inexplicable way, guided by a universal unifying force we cannot yet comprehend. If not God, then perhaps we, ourselves, have the power to bring like and like together. Are coincidences, perhaps, a glimpse of our latent psychic powers, akin to telepathy, clairvoyance, and premonition?

Our fascination with both coincidence and the paranormal come together neatly in our passion for horoscopes. Even people who profess to be devout skeptics have been spotted surreptitiously checking their horoscopes.

It began thousands of years ago when our distant ancestors failed to understand that a solar eclipse—during which day turned dramatically into night—was nothing more than the coincidental alignment of a ball of gas and a ball of rock.

These events also, inevitably, coincided with events on Earth. Chroniclers down the ages have recorded how eclipses and planetary conjunctions have coincided with famines, earthquakes, volcanic eruptions, major military defeats, or victories and the deaths of emperors and kings.

Fascination with these coincidences eventually formalized into the prediction business. Most famously, the sixteenth-century French astrologer and physician Nostradamus translated his study of the stars and horoscopes into a catalogue of dramatic, if inscrutable, prophecies. Some credit him with anticipating the French Revolution and the First World War. More recent claims that Nostradamus accurately foretold the September 11 terrorist attacks on the twin towers of the World Trade Center in New York have been exposed as an urban myth.

> In 1987, journalist and astrologer Dennis Elwell hit the headlines after he warned of a possible disaster at sea—just days before 188 people died when a car ferry, the Herald of Free Enterprise capsized off the coast of Holland.
>
> Elwell explains the astrological evidence that prompted his warning. "Technically the March 1987 solar eclipse was raising the temperature of a square between Jupiter and Neptune, planets which, when working together, indicate both sea travel and big ships. Eclipses bring the matters signified into high profile, and tend to be associated with misfortunes, although positive outcomes are also possible."
>
> Elwell sent identical letters to two shipping companies, alerting them to the potential hazard. The letter said, "The emphasis is on the sudden and disruptive. While I am not in the prediction business, it would be no surprise to find that, at the very least, sailing schedules were upset for some unexpected

reason. But there has to be a possibility of rather more dramatic eventualities, such as explosions."

Only nine days after the car ferry company replied that their procedures were designed "to deal with the unexpected from whatever quarter," their ship, the Herald of Free Enterprise, capsized.

Elwell's prediction was dramatically and tragically accurate. But was this just coincidence? We never hear about all the psychic predictions that turn out to be wrong, the foretold disasters that stubbornly fail to happen. Perhaps there aren't any; though that seems a little unlikely. Perhaps the mistakes are quietly swept under the carpet. And how many amazing predictions are only revealed after the events they heralded? Predictions by hindsight!

Less spectacular predictions pour out of our newspapers and magazines every day in the horoscopes. But how likely is it that the celebrity astrologers responsible will be able to anticipate fortune or misfortune in our lives?

Whether we "believe" in astrology or not, most of us can take pleasure from horoscopes. When the predictions appear to come true, it is hard not to pause and wonder.

On August 25, 2003, three different national newspaper horoscopes offered a variety of advice for people born between March 21 and April 19 under the sign of Aries. The first promised the arrival of long-awaited money, a new kind of inner strength that would help with "love choices" and the solution of a family mystery; the second predicted the discovery and unleashing of "real hidden power" that would open up wonderful possibilities; and the third advised that the alignment of Jupiter and Uranus could force changes regarding a commitment that had become a burden. He warned, "You're letting imaginary fears force you to try so hard to make everything perfect that there's no time for things you like best. Do something about it."

What does it all mean? And what were Arians to conclude if any of those predictions came to pass? Those who believe in the prophetic power of horoscopes use them to guide themselves through crises in their lives. Others dismiss any apparent correlations between prediction and events as simple coincidence. Should we dismiss accurate predictions as the product of pure chance or is something more interesting going on? Are our lives already written down for us in the stars? Is there a template for our lives in the planets?

Our historical fascination with horoscopes would be legitimate if it were possible to prove scientifically that from the moment of birth our lives are bound inextricably with the movement and interactions of the planets and, therefore, that coincidences between predictions and subsequent events are meaningful. Astrologers say they are, but then it pays their wages. What about an astrologer turned scientist? Pat Harris is running a research project at Southampton University looking at, among other things, the possible impact of the planets on pregnancy and childbirth. She stresses that she doesn't "believe" in astrology, she is simply interested in studying it scientifically to establish whether coincidences associated with the juxtaposition of the planets can be attributed to anything other than pure chance.

After studying the star signs of a number of mothers-to-be she is able to say that there is a strong correlation between the influence of Jupiter and successful pregnancies and healthy births.

But how could Jupiter cause the successful birth of a baby?

"I can't say that it does. At this stage we can only talk about correlation—or synchronicity, as Jung would call it. When something goes on in the heavens, something goes on down on Earth. They appear to be connected, but we don't know if one causes the other."

Astrophysicist Peter Seymour has gone further, and has attempted to come up with a theory of how the planets might have a physical impact on human destiny.

Seymour sees the solar system as an intricate web of magnetic fields and resonances. The Sun, Moon, and planets transmit their effects to us via magnetic signals. Magnetism, he points out, is known to affect the biological cycles of numerous creatures here on Earth, including humans.

The planets, he suggests, raise tides in the gases of the Sun, creating sunspots. Particle emissions then travel across interplanetary space, striking the Earth's magnetosphere, ringing it like a bell. He believes the various magnetic signals are then perceived by the neural network of the fetus inside the mother's womb, "heralding the child's birth."

French psychologist Michel Gauquelin devoted his life to trying to find out if there was a scientific basis for astrology. He conducted major studies exploring statistical links between the births of eminent doctors or politicians or soldiers and particular conjunctions of the planets. He discovered, for example, that an unlikely percentage of French professors of medicine had been born when Mars and Saturn were dominant. Mars was also shown to be particularly significant in the birth charts of more than two thousand leading athletes.

He found many other similar correlations:

Sportsmen	Mars, lack of Moon
Military	Mars or Jupiter
Actors	Jupiter
Doctors	Mars or Saturn, lack of Jupiter
Politicians	Moon or Jupiter
Executives	Mars or Jupiter
Scientists	Mars or Saturn, lack of Jupiter
Writers	Moon, lack of Mars or Saturn
Journalists	Jupiter, lack of Saturn
Playwrights	Jupiter
Painters	Venus, lack of Mars or Saturn
Musicians	Venus, lack of Mars

By no means did all of Gauquelin's findings came out in support of astrology. His early work on zodiacal signs found no evidence to support the astrologers' claims. Throughout his life he faced accusations from the scientific community that his findings were inaccurate or even fraudulent. In 1991 he committed suicide, after first destroying much of his original data.

A piece of more recent investigative research has thrown up a possible astral link between car thieves and their victims. Using statistics provided by police, it has been discovered that car thieves and the owners of the cars they steal commonly share the same birth sign. The inference is that if you are born under the same sign, you share similar preferences, including your taste in motor vehicles. It's doubtful how much comfort that will offer the freshly bereaved former owner of a shiny new Porsche Carrera GT. What will the thief come after next? His wife?

> *Amateur astronomer Peter Anderson regarded astrology as "bunkum." One day he found a newspaper lying on a desk, opened by chance at the horoscopes and, despite his innate skepticism, found himself glancing at the prediction for his birth sign—Capricorn. It said he would be offered two jobs in the next week. He had a good laugh about it. The next day he was offered two jobs. . . .*

The more coincidences we observe in our lives, the more they excite and entertain us. And strange, inexplicable coincidences, predicted or unpredicted, are happening around us much more frequently than we realize. We tend only to notice those events that are brought to our attention or are so startling that we can't miss them.

Writer, heroin addict, and wife-killer William S. Burroughs believed that our paths through life are littered with coincidences and that all of them are significant. He kept records of his dreams, a scrapbook of newspaper clippings, and notes on seemingly anything that occurred, searching for the coincidences in his life—and their meaning. Burroughs suggested that we should all increase our awareness by being more observant. He advised friends to take a walk around the block, come back and write down precisely what happened, with particular attention to what they were thinking when they noticed a street sign, or passing car, or stranger, or whatever caught their attention. He predicted that they would observe that what they were thinking just before they saw a sign, for example, would relate to that sign. "The sign may even complete a sentence in your mind. You are getting messages. Everything is talking to you," he told them. "From my point of view there is no such thing as coincidence."

The mystery of coincidence is seductive. We want to see coincidences around us—we need to see them. But our enthusiasm can lead us astray, resulting in false sightings. A sense of missionary zeal may have been burning in the breast of the person who first noticed, and posted on the Internet, the "extraordinary coincidences" between the classic Judy Garland film, *The Wizard of Oz*, and the bestselling rock album, *Dark Side of the Moon*, by Pink Floyd. All you have to do, explained the Web site, is play the album at the same time as the film to spot the amazing synchronicity. It suggested that some sort of cosmic force had been at work here, unifying the creative output of the musicians and the moviemakers.

Such was the Internet-surfing public's fascination with this revelation that shortly after this "coincidence" theory was first described on the Internet, sales of the Pink Floyd album surged and copies of *The Wizard of Oz* were swiped from the shelves of film rental shops.

To check out these coincidences for yourself you have to carefully follow some very precise instructions:

First of all, buy your album and video. Start your Pink Floyd music at the precise moment the MGM lion finishes its third and final roar . . . "and you will find some very interesting coincidences."

You'll know you've got it right when the first chord of the song "Breathe" comes in at the same time as "produced by Mervyn LeRoy" is displayed on the screen, and Dorothy is teetering on the fence to the pigpen when the band is singing "balanced on the biggest wave."

Dorothy falls into the pigpen just as you hear the words "race towards an early grave," and the music changes at the same time. Dorothy holds a little chick up to her face in a caring manner as you hear the band sing the words "don't be afraid to care." When the band sings "smiles you'll give and tears you'll cry" the Lion and the Tin Man are smiling and the Scarecrow is crying. The song "Brain Damage" begins at about the same time the Scarecrow starts singing "If I only had a brain." When the Munchkins are dancing after Dorothy arrives in Oz, the scene appears to be perfectly choreographed with the song "Us and Them."

And on and on it goes in similar fashion.

Now for those who haven't immediately rushed off to try this experiment, you should be warned that you may find the experience a little frustrating—a complicated and unsatisfactory way to spend an hour or so of your life. Does it confirm that the world is full of wonderful and exotic and inexplicable coincidence? What do you think, Toto?

But the very fact that—at a conservative estimate—thousands of people have gone to the trouble to explore this alleged phenomenon says a lot about our collective need as a species to find meaningful coincidences in our lives.

Our love of coincidence seems to be inextricably tied to that other fundamental human need—to understand the meaning of existence. In both cases we seem desperate to convince ourselves that there is more to it, more to coincidence and to life, than random chance and serendipity.

Douglas Adams bent his mind to the meaning of life in his hugely popular *Hitchhiker's Guide to the Galaxy* series. The comic novels famously come to the conclusion that the answer to the meaning of life, the universe, and everything is 42. Adams might have found it harder to explain the coincidence that occurred to him at Cambridge railway station.

> *Adams went into the station cafeteria and bought a packet of cookies along with a newspaper, then sat at a table. A stranger sat down, opened the bag of cookies and started to eat them. There was obvious confusion over the cookies' ownership. "I did what any red-blooded Englishman would do," says Adams. "I ignored it." Both men alternately removed cookies from the bag until it was empty. It was only when the stranger left that Adams realized he had placed his newspaper over his own identical bag of cookies. "Somewhere in England there is now another man telling the exact same story," Adams observed, "except that he doesn't know the punch line."*

Perfect coincidences make perfect stories—better than anything even an accomplished storyteller like Douglas Adams could have invented.

Few of us encounter the "double cookies packet" coincidence

scenario. More common is the experience of unexpectedly bumping into your brother-in-law at a nudist camp. That happens a lot. But wherever an experience appears on the Richter scale of coincidence, it always feels amazing and very special to the person involved. The universe has singled us out for special attention. "Look what I can do," it seems to be saying.

Our love of coincidence knows no bounds. As if there weren't enough naturally occurring coincidences to satisfy our insatiable need, we fashion them ourselves, working them elaborately into our art and literature. We are in love with the shape and sound, the rhythm and rhyme of coincidence. It makes us laugh, too:

> *A baby goat wanders too far from his nanny, tumbles over a cliff, and is swallowed by a large flat fish. A fisherman hauls in the flat fish, cuts it open and out jumps the baby goat. The fisherman says: "What's a nice kid like you doing in a plaice like this?"*

Why does this joke work? Because we enjoy the coincidence of these ideas coming together. We love the surprise of seeing or hearing words used in unexpected ways. Puns are coincidences of sound and meaning—two parallel thoughts tied together by an acoustic knot. The English language, cobbled together as it is from Latin, Norse, and French, as well as your common or garden Anglo-Saxon, has innumerable words of similar or identical meaning—as well as many words that sound the same but mean something completely different. It makes wordplay a cinch. For example:

> *Time flies like an arrow.*
> *Fruit flies like a banana.*

Coincidence is integral to the interior creative process. Every second our minds make hundreds of links, rejecting most, bringing together unlike elements to see if, in sum, they make more than their separate parts. It is this meaningful linking from which an idea is born, be it a pun, an apt metaphor, a rhyme, or a relentless plot mechanism.

Secretly we are delighted that an article in a urology journal should have been written—and this is true—by J.W. Splatt and D. Weedon, even though we maintain in public that we are above such childish amusement. The coincidence that links Messrs. Splatt and Weedon to their profession is by no means unique. Just ask Cardinal Sin of Manila or the former Saudi Arabian oil minister Sheik Yamani. There are many more examples.

> *A police officer Neil Cremen, who luxuriates (for reasons unknown) under the nickname "the Dog," was sent to investigate a complaint about a savage dog. When he arrived at the house, the dog duly bit "the Dog." In the hospital Cremen was treated by a Dr. Bassett. The owner was taken to court and prosecuted by a lawyer named Barker.*

There is immense satisfaction in the sound of those references closing together. There's completeness about such a story, a relief from chaos: pattern resolved, cycle fulfilled, ends tidied away. Is this what motivates us? Neatness?

Coincidences, real or artfully contrived, can make us laugh out loud. Here's a true story:

> *A woman sunbathing on an inflatable lobster was washed out to sea. She was rescued by a man on a set of inflatable false teeth.*

But this one's made up:

> *An old man goes to heaven, and sitting at the reception desk in heaven is Jesus, who calls the old man forward and says, "Old man, welcome to heaven. I have to take some details—could you tell me your name?" The old man says, "My name is Joseph." And Jesus says, "Well there's a coincidence, when I was on earth my father's name was Joseph." And the old man says, "Well I had a little boy, you know, he'd be about your age by now." And Jesus says,*

"Well how extraordinary . . . but I left home when I was quite young." And the old man says, "Yes, my little boy left home when he was young. He went away with his friends, they got involved in magic and other mystical stuff." And Jesus says, "Another coincidence—how extraordinary, that's exactly what happened to me. Tell me, what was your job back on Earth?" And the old man says, "I was a carpenter." And Jesus says, "That's an amazing coincidence, that was my father's job too . . . you don't think that you and I could be . . ." And the old man says, "No, you see, my little boy was not born like ordinary boys." And Jesus says, "That's how it was with me." And the old man says, "Look, I would know my little boy anywhere, you see he has these little holes in his hands and feet." And Jesus says ". . . you mean like THIS." And the old man says, "I can't believe it." And Jesus says, "You must believe it—so many coincidences, you must be my earthly father Joseph." And the old man says, ". . . and you must be my little boy—Pinocchio!"

One of the most memorable images from the history of silent movies is of Buster Keaton standing innocently in front of the collapsing façade of a house. The house falls, but by a stroke of luck Keaton happens to be standing directly in line with an open window. Our delight in this apparent coincidence mirrors the satisfaction we get from true dramas—like the woman saved from serious injury when her chimney crashed through the ceiling, smashing the end of her bed, seconds after she had curled up her legs, or the soldier saved from a bullet fired at his heart by the silver locket containing a picture of his fiancée.

Coincidence is the driving force behind the most popular of all modern forms of entertainment. Where would *As the World Turns* and *All My Children* be without all the outrageous coincidences that propel their convoluted and incredible storylines—the amazing chance encounters and complex interrelationships. Hitherto unheard-of brothers, sisters, uncles, mothers, and sons materialize from thin air, just in time to resolve tricky plot dilemmas or revive flagging ratings. The more far-fetched the coincidence, the more we love it.

But, as usual, truth tends to be stranger than fiction. Would modern soap opera fans swallow a storyline that imitated the real-life relationships revealed by the marriage announcement carried in an American newspaper back in 1831?

> At Saco, Maine, on Christmas Eve, by the Rev. William Jenkins, Mr. Thophilus Hutcheson to Miss Martha Wells; Mr. Richard Hutcheson to Miss Eliza Wells; Mr. Thomas Hutcheson to Miss Sarah Ann Wells; Mr. Titus Hutcheson to Miss Mary Wells; Mr. Jonathan Hutcheson to Miss Judith Wells; Mr. Ebenezer Hutcheson to Miss Virginia Wells, and Mr. John Hutcheson to Miss Peggy Wells.

Another popular dramatic device that also materializes in real life and relies entirely on coincidence is mistaken identity.

There are more than six billion people in the world. Most of them have a similar arrangement of arms, legs, and facial features. They come in a limited range of colors, heights, and builds. That two people might look very similar is perhaps not so extraordinary, yet we take great delight in the coincidence when they do. Actors and actresses have built careers on the fact that they have a more than passing resemblance to the president or Michael Jackson, or Britney Spears. Impressionists like Dana Carvey delight us with their capacity to both look and sound like a range of familiar figures.

We experience a strange and slightly disturbing frisson when friends say they have met someone who is the spitting image of ourselves. "He had wild staring eyes, a bad haircut, and poor posture," we're told. "I was certain it was you."

Alfred Hitchcock's film *The Wrong Man* draws on the real-life drama of Manny Balestrero—a New York Stork Club musician falsely accused of a series of holdups. He was arrested and charged in 1953 after several witnesses identified the bass player as the culprit. He was eventually released when his "double" was apprehended.

The idea of a double or "doppelgänger" is common throughout literature and in drama. The Hollywood film, *The Talented Mr. Ripley,* is based on the idea that someone's life could be completely taken over by a double, a theme that fiction writers have warmed to over the years.

In some specific circumstances, obsession with this particular kind of coincidence can actually have a physiological cause—seated deep within the brain. Swiss neurobiologist, Dr. Peter Brugger of Zurich University Hospital, has evidence of a brain condition that can make the sufferers believe they have a real double. This rare delusional disorder is called Doppelgänger Syndrome.

Sufferers imagine they see an exact replica of themselves. In extreme cases the sensation is accompanied by a belief that they are gradually being replaced by their doppelgänger. In one of Dr. Brugger's cases a man felt so persecuted by his double that he shot himself to get rid of it.

Within the world of fiction, this classic doppelgänger scenario features in Dostoevsky's *The Double*, Oscar Wilde's *The Picture of Dorian Gray*, and a number of other works. Dr. Brugger speculates that some authors who have used this particular dramatic device may themselves have been suffering from Doppelgänger Syndrome and writing from experience.

Coincidence has been an engine of literature for centuries, its exponents including no lesser wordsmiths than Shakespeare and Dickens—who seldom hesitated to introduce extraordinary chance events to keep their plots bubbling along.

Dickens' *A Tale of Two Cities*, for example, revolves around successive coincidences, that feuding central characters Sydney Carton and Charles Darnay look alike, resulting in Carton going to the guillotine in Darnay's place, and that Darnay, who marries the heroine, Lucie Manette, happens to be the nephew of the marquis who had her father imprisoned.

Shakespeare's *Twelfth Night* is packed with coincidence. Viola believes her twin brother Sebastian has died. To protect herself in a strange country, Viola pretends that she's a boy—Cesario. But in her boys' clothes she looks just like Sebastian, and when he turns up with his new friend Antonio, everyone thinks that he is Cesario. The beautiful Countess Olivia has fallen in love with Cesario (not realizing *he* is Viola) while Cesario has been taking messages of love from Duke Orsino to Olivia. Eventually all is resolved in a happy, if somewhat coincidence-packed, ending.

Writer and literary critic John Walsh believes *Macbeth* to be the Shakespearean play that makes the most interesting use of coincidence. Macbeth is assured by the three witches that he should fear no man born of woman...and certainly not before Birnam Wood comes to Dunsinane. So no worries there for Macbeth. Unfortunately it transpires that his adversary Macduff was born by caesarean section. Macduff's soldiers then creep up on Macbeth at Dunsinane by using the branches of the trees of Birnam Wood as camouflage.

Says John Walsh, "I imagine Macbeth turning away, smacking his forehead with the palm of his hand and saying, 'How unlikely a coincidence is this?!' Just before his head is severed by Macduff."

It has been said of Charlotte Brontë's somewhat romantic plot developments in *Jane Eyre* that she "stretched the long arm of coincidence to the point of dislocation." The same could be said of countless writers over the centuries. Not that this has made them any less successful. Quite the contrary.

Coincidence is more sparingly used in contemporary fiction. The more sophisticated modern reader can be left feeling short-changed or even cheated by its overuse. Not that this much inhibited James Redfield, author of *The Celestine Prophecy*. Despite its modest literary credibility, the book became a number-one bestseller around the world. Its plot revolves around the search for the "nine key insights to life itself." The first insight, it transpires, occurs when we become "fully conscious of the coincidences in our lives." This turns out to be little more than the justification for one of the most unconvincing, coincidence-laden plots ever conceived. Still, the book-buying public of the world can't be wrong. Our love of coincidence knows no bounds.

American author Paul Auster is an enthusiastic modern advocate of coincidence as a structural or narrative device. He says, "We are continually shaped by the forces of coincidence. Our lifelong certainties about the world can be demolished in a single second. People who do not like my work say the connections seem too arbitrary. But that is how life is."

And that's how it was for British crime writer Ian Rankin— author of the Inspector Rebus crime novels—when he was invited to contribute to an obituary of novelist Anthony Powell who had just died. Powell was the author of the twelve-volume epic, *A Dance to the Music of Time*, which is packed with coincidence and synchronicity— school friends bump into one another after forty years; somebody will be thinking about a painting just before being introduced to the artist.

Rankin had been introduced to the books by a friend at university who gave him the first three volumes as a birthday gift. Powell's death prompted him to start reading the books again. He took the first couple of volumes with him on a trip to Harrogate for the annual conference of the Crime Writers' Association.

"Coincidence," he says, "has dogged my writing career." He published a novel about a "magic circle" of judges and lawyers, only to have the police in Edinburgh investigate a similar claim two years later. A year later, sitting in the south of France, he dreamed up a story idea about an alleged war criminal living quietly in Edinburgh, only to find that Scottish Television was making a documentary about a real war criminal living in the city.

"More recently, I encountered a gentleman with the surname Rebus who lives in Edinburgh's Rankin Drive, and a police officer with the exact same job title and surname as another character in one of my books."

At the Harrogate conference one speaker showed a slide of a truck that had lost control and smashed into the bedroom of a bungalow. The owner of the bungalow was sick that day and confined to bed. But the telephone had roused him from bed seconds before the truck toppled a wall onto the spot where he had been. The telephone call had been a wrong number.

Rankin returned home on the Sunday evening and collapsed onto the sofa, reaching for the TV remote. As he channelsurfed, he saw a face he recognized. It was one of the contestants on *Who Wants to Be a Millionaire?* "It was Alistair, an old friend—and he'd just won $225,000."

It wasn't until Rankin was going to bed that he placed the final piece into the coincidence jigsaw. He remembered what else his friend had done. Years earlier, when they had been at university together, Alistair had given him a birthday present of three Anthony Powell books. . . .

So why do we love coincidence? Is it because we are intuitively celebrating what Arthur Koestler called a universal principle that "things like to happen together"?

Perhaps coincidence is fundamental to the human condition. We crave and need the patterns and rhythms and symmetry it provides. It brings respite from disorder. And perhaps our brains are hard-wired to both seek out and create synchronicity. We absolutely need doppelgängers and parallel universes in which alternative versions of ourselves live—only more successfully, of course.

Mathematician Ian Stewart has studied coincidence. His scientist's view of why we love it is more prosaic. "Well, it gives us great stories to tell at the bar," he says.

IT'S A SMALL WORLD—
COINCIDENCE AND
CULTURE

The oldest observed and most enchanting coincidence of nature has led wise men and children a merry dance down the ages to many and various inventive explanations. It is the fact that the Sun and the Moon appear equal in size in our sky. We know now it's all a matter of perspective, but that's only because clever people have told us so.

The first clever people had very little reliable knowledge to build upon. The 6th-century-BC Greek philosopher Heraclitus estimated the Sun to be a foot in diameter. This would have made its distance from Earth around forty-four yards. It's easy to say now, with the help of five minutes' research on the Internet, that Heraclitus was wrong. We can even give him the figures: the Sun is 830,247 miles across, compared with 2085 miles for the Moon. The Sun's diameter is four hundred times larger than the Moon's. The Sun is also four hundred times farther away from us than the Moon. It is this relative distance that makes the two bodies equal in size in our perception.

Given the apparent randomness of the cosmos and the vast distances involved, it is a truly remarkable coincidence that from our unique perspective the Sun and the Moon should appear the same. But coincidence is all it is, however potent the Sun and Moon's symbolism in our lives and folklore as a complementary pair of equal opposites.

What was magical in the past, however, doesn't necessarily hold the same mystery for us now. On the other hand we are not immune to magical interpretations of newly perceived coincidences. In fact the evidence suggests that our tendency to opt for paranormal explanations is increasing. One reason is that we experience a lot more coincidences than people did in the past, and the frequency multiplies every year.

Our ancestors lived in smaller communities than ours, traveled less frequently, less far, and were exposed to a narrower range of experiences. Opportunities for unlikely correlations in their lives were more limited. They made the most of those that came their way, often investing them with profound significance.

Ostensibly the modern world is less superstitious, yet it is also a place in which seeming magic is more likely to happen. It's a busy and bewildering place, growing ever busier and more bewildering. Within the last hundred years human society has accommodated several dynamic technological revolutions, each of which has transformed the pace and scope of individual experience. We now have mass mobility, mass communication, and mass access to computing power; and we have that inexhaustible information regurgitator, the Internet.

The solemn maxim "know thyself," written above the temple of the Ancient Greek oracle at Delphi, may be—as it ever was—more honored in the breach than in the observance, but now at least we seem to know everything else. There are billions of items of computer-sorted information at everybody's fingertips, broadening our view but not necessarily our understanding.

Profligacy of information makes the possibility of coincidence more likely. The statistician's law of large numbers states that if the sample is very large even extremely unlikely things become likely.

Well, the sample base we expose ourselves to every time we travel abroad or log on to the Internet is vast. "It's a small world!" we exclaim, as the correlations come together. One thing is certain: the wider the World Wide Web, the smaller the world. Today we are wired for coincidence.

But while our experience of coincidence has increased, our knowledge of probability hasn't kept pace. Most of us have a better grasp of elementary mathematics than did the average American colonist, but the sheer volume and complexity of our experience of coincidence makes it harder than ever to sort out the fantastic from the mathematically feasible.

That's why the most consistent factor of reported coincidences is the insistence by their observers that they aren't coincidences at all. They are brought about by angels, or magic, or sock goblins, or space aliens playing around with the postal services—anything but simple chance.

The problem is, chance isn't simple. You need to know a fair bit of mathematics to be able to work out probabilities. Scientists and bookies do it inside their cool heads but most of the rest of us are dismayed by the arithmetical effort and rely instead on intuition, which is demonstrably bad at estimating probability. Human beings are very easily impressed. What seems utterly unlikely to a human being often turns out to be extremely probable in the cosmic scheme of things. Think Sun and Moon.

Or think Bible codes. According to some ancient accounts, the Book of Genesis in the Hebrew Bible, which is said to have been dictated by God himself, contains codes that if deciphered will reveal many other messages for mankind. It has long been a respectable pursuit for scholars belonging to remote and dusty religious orders to attempt to detect hidden patterns therein, discounting the spaces and punctuation in the text and treating the letters as a regular matrix. Inevitably, given the number of letters the Bible contains, and the fact that written Hebrew contains no vowels, many coincidental word patterns have manifested themselves in these searches, to which significant meaning has been attributed.

Computer science, far from making this arcane procedure seem even more eccentric, kicked the whole code detection business into a different league by increasing the speed and the variety of ways in which the letter matrix could be analyzed. Words could be identified running forward, backward, vertically, and diagonally in the text. Using a procedure called equivalent letter spacing it could also find words consisting of letters that were not adjacent, but were spread out in the text, each letter separated by the same number of nonrelevant letters. Computer searches carried out by the prominent Israeli mathematician Professor Eliyahu Rips discovered incredible examples of conceptually related words adjacent to each other in the text, such as the names and birthplaces of famous rabbis. The discovery of the name of the murdered Israeli president Yitzhak Rabin next to a reference to death and a vertical "Kennedy" running through the phrase "assassin that will assassinate" seemed to suggest a prophetic quality.

Skeptics were slow to counter the claims of the researchers and Michael Drosnin's book about the phenomenon, *The Bible Code*, sold millions. Even today it seems few things excite us more than the prospect of proof of the paranormal. It took time for other teams of statisticians to find conceptual flaws in Professor Rips's painstakingly rigorous experiments. Meanwhile Brendan McKay, professor of computer science at the Australian National University, used Rips's system to find prophetic correlations of death and murdered presidents in *Moby-Dick*. In the end the Bible codes merely demonstrated that given enough letters, coincidental word patterns will emerge, and that many of them, given a little interpretation and a great deal of excitement, will appear to have meaning.

In 1967, sociologist Stanley Milgram predicted that there were only six degrees of separation between any two people on the planet. The idea entered dinner party folklore, but few people realized that Milgram's attempts to prove it were unsuccessful. Recently, however, another sociologist, Duncan J. Watts, successfully proved a similar proposition. Watts assigned sixty thousand people a target person, possibly living in a different country and certainly from a different walk of life, and instructed them to attempt to pass an e-mail message to that person by forwarding it only to someone they knew, with a request to forward it on in the same way. On average it took between five and seven e-mails to hit the target.

Watts's experiment is an effective demonstration of the smallness of the world, but when incredible-seeming correlations happen to us outside of a scientific experiment they feel eerie and magical. There are many reasons why we might *want* them to be magical. The best is summed up by Richard Dawkins, a scientist famous for debunking the paranormal, who says, rather generously, given his standpoint, that we have a "natural and laudable appetite for wonder."

That appetite for wonder rekindled Joyce Simpson's faith in God. Joyce, of DeKalb County, Georgia, saw a sign in May 1991 that changed her life. To everyone else it was a Pizza Hut advertisement, but Joyce, who at the time was disillusioned enough with religion to be considering quitting her church choir, saw only salvation. Shining forth from a forkful of spaghetti was the face of Jesus.

A skeptic will say that if you look closely enough and with enough emotional motive you can see the face of God in any picture—spaghetti, chicken nuggets, deviled eggs—but there'd be no point telling Joyce Simpson that, nor the rapturous Georgians who lined up in their cars to see the billboard miracle for themselves once the news got out.

Wonder engages the emotions. Wonder changes lives. Wonder makes you sit down and write that long postponed letter to Herbert Krantzer. "Dear Herbert. You won't believe this. I was washing my old VW Beetle, prior to finally selling the damn thing, and I found a note from you, dated June 1986, inside one of the hubcaps, wishing me a 'long and happy journey!' You must have slipped it in there on my wedding day all those years ago. . . ." It makes you feel especially blessed when Herbert writes back to say how good it was to receive your letter, particularly at that moment, for he is in the very process of finding an old VW Beetle for his son.

What made you take the hubcap off, for the first time in seventeen years, at that particular moment? To clean it, the skeptic will say.

The letter finding event was acausal; in other words, you didn't find it *because* it might be propitious to contact an old friend at that particular point in time.

To most human beings this is a pretty bland interpretation. It's the word "acausal" that rankles. It renders what seemed full of wonder flat and dull. The letter writer may consider himself to be a rational thinker, but his mind is more interested in indulging the possibility that a guardian angel is smiling on him, or that his relationship with Herbert is so significant that some kind of telepathy is in play. He would rather take the agnostic "who knows?" position than consign it to mere chance.

Acausality doesn't do justice to the experience, especially when the experience is very personal. If a man dreams one night that his friend Moriarty is dying and then wakes up to be told that Moriarty has actually died, the notion that he may be psychic is extremely hard to resist. As is the notion that God sent him a warning to soften the blow, or that there are parallel universes in different time dimensions to which he might suddenly, because of the strength of his emotional compact with his friend, have gained access, or that the emotional right-hand side of the brain, containing a primitive intuitive consciousness suppressed by centuries of evolution, has woken up during his sleep, or that an event has come about because of the motivating power of his own thought process (on second thoughts, scrub that one). . . . There is no shortage of such explanations and every one of them is more interesting than arbitrary, impersonal, *acausal* chance. What makes them utterly irresistible is the way they engage the dreamer emotionally with the fact of the death: they impart a sense of having been present, or somehow consulted, at the end.

Statistician Christopher Scott has worked out the odds in the UK of dreaming of a friend's death the night it happens. Basing his calculation on fifty-five million people living an average of seventy years and experiencing one friend's death dream per lifetime, and then factoring in a national death rate of two thousand every twenty-four hours, Scott reckons there'll be an accurate death dream in Britain about every two weeks. It's human nature to recall only interesting stories, so accurate dreams are widely reported and frequently retold, while millions of dreams about dying friends who turn out the next day to be on the mend are routinely discarded from memory.

"How do you know," the dreamer might say, "that all the dreams that didn't come true weren't of an inferior quality to mine? My dream about Moriarty had *authority*. It was so vivid it had to be true. It was as though the gods were intervening in human affairs—it was a *deus ex machina!*"

"Why would Zeus have confided in you?" asks the skeptic.

"Well, you know . . . we're pretty close."

"What about that time you dreamed you were being chased naked through Starbucks by Julia Child waving a cleaver? Has that come to pass?"

"Not exactly . . . though it *could*, of course. Anyway, not *all* my dreams are prophetic."

"How many have been?"

"Well, there's the dream about Moriarty . . . and I once dreamed I was going on a long journey and soon after I won a weekend trip to Paris . . ."

Out comes the calculator again. "So . . . two prophetically accurate dreams in thirty-six years . . . let's say you have three dreams a night . . . that's one clairvoyant vision in every 19,710 dreams, or put another way . . ."

"No you're wrong, Clever Dick, because there's something I haven't told you. None of this—Moriarty, the dream, your calculator, the whole of existence and every skeptic who ever set foot in it—actually exists. The universe is actually a figment of my imagination. I made it all up! In fact I'm making you up right now. Work that out on your calculator!"

You see how coincidence sparks the imagination! Note also that all explanations except chance grant the observer a starring role. Nowadays most of us tend to accept the skeptic's rationale, at least outwardly, but privately we like to at least flirt with the center-stage glory coincidence gives us. It's a natural enough desire that goes hand in glove with the need for an explanation for the universe that makes us feel less like a speck of random space dust and more like a cosmic player. Even the most skeptical probability mathematician, on finding a bottle washed up on a beach in Madagascar containing a note addressed to him, might be tempted to entertain that awesome possibility.

Yet if this event is meaningful, what exactly does it mean? He can only pitch a guess at that. Or he can seek the advice of a New Age counselor or shaman, if he can find one who is trustworthy and reliable and not given to irrational flights of fantasy.... The realization at this point that he is stepping into a supernatural rocket ship fueled by high-octane superstition and without a qualified pilot might remind him that he is a skeptical probability mathematician. Those who abandon empirically testable evidence as a basis for important life choices in favor of the subjective interpretation of random events, follow a convoluted and perilous path, as history has demonstrated only too often.

Two and a half thousand years ago, when Sophocles wrote *Oedipus Rex*, nobody was backward about predicting the future. They could research their destiny as readily as we can research our history and everyone had a direct line to the gods. Clotho, Lachesis, and Atropos sound like members of the Marx Brothers, but to the average man they were very real, and not a bit funny. They were the three Fates, the indifferent celestial beings that meted out the thread of life apportioned to each mortal, decided on a few choice life qualities (tragedy, illness, etc.), and efficiently snipped it off at the due date. This thread was a man's *moira* (allotment). He couldn't erase the best-before date, nor could he escape his *moira*'s negative elements, though he could, if he was foolish, make things much worse for himself.

Comets and other natural phenomena were obvious augurs; not cosmic coincidences these, but harbingers of specific events on Earth, usually disasters. The fall of Jerusalem, the death of Julius Caesar, and the defeat of the English by William the Conqueror were all said to have been augured by comets. King Harold's defeat was mapped out by Halley's comet. It came around again in 1986, presaging what evil this time? The explosion of the space shuttle *Challenger*? The assassination of Swedish prime minister Olof Palme? The invasion of the United States by Crocodile Dundee?

It's unusual to make such associations today, yet a minority of people still do. It's not hard to find clairvoyant sites on the Internet asserting cast-iron connections between historical events and comet appearances that preceded them. The astronomer Carl Sagan, who waged a war against "baloney and pseudoscience," said that since human history is intrinsically unhappy, "any comet at any time, viewed from anywhere on Earth is assured of some tragedy for which it can be held accountable."

In the time of Sophocles, such things as comets seemed more authoritative, and the oracle was recognized as providing an insight into the Fates' deliberations, a sort of advance preview of your life. Just how dangerous reliance on such a system could be is demonstrated in many stories of the Delphic oracle, in particular that of poor Oedipus who, if we believe that his fate was preordained, as Sophocles appears to have done, was doomed before he was even born. It wasn't enough that each tragedy Oedipus had to endure had been written in the stars, it was also pointed out to him before it happened by well-wishing clairvoyants! Wriggle though he and his family might to escape their fate, nothing they did could prevent the predictions from acting themselves out. In fact it was the very act of wriggling, also foreseen by the cunning immortals, that set in motion the events that had been predicted.

To a modern skeptic, Oedipus's tale is nothing to do with fate; it's all about coincidence. It's just very, very bad luck. Coincidences do tend to cluster—any statistician will tell you that—and Oedipus had the misfortune to be the point where all the bad luck congregated that particular eon.

Oedipus's story is as predictable as an episode of *The OC* and almost as depressing. His father, King Laius of Thebes, was told by the oracle that the boy would grow up to murder his father and marry his mother.

It was an unusually unequivocal response from the Pythia, the Delphic priestess who, in mystic trances, garbled the prophecies.

Normally the Pythia's pronouncements took the form of riddles that lent themselves to more than one interpretation, a common enough play of clairvoyants, astrologers, and psychics both then and now (the blanket prediction enables them to say "I told you so" whatever the outcome). When King Croesus of Lydia was contemplating making war with Persia in 550 BC, he sent emissaries to the Pythia with lavish gifts of gold and silver, three hundred cattle and a gold bowl weighing a quarter of a ton. They were told, "You will destroy a great empire." Pleased with the prediction, Croesus attacked Persia and destroyed a great empire—his own.

There was no such leeway for interpretation in King Laius's oracle. No matter what slant he tried to put on it, the future was already pretty bad. Laius tried to sidestep fate by instructing that the baby should be left on the slopes of Mount Cithaeron to die. The baby was saved by a shepherd and taken to Corinth, where by an incredible coincidence it was noticed and adopted by the childless Corinthian king, Polybus. Oedipus grew into a golden child, but he, too, consulted the oracle, which glibly repeated that he would kill his father and marry his mother.

Believing Polybus to be his father, Oedipus left Corinth at once for Thebes. On the way he met his real father at a crossroads, fell into an argument with him, and, in one of the few road rage incidents recorded in the ancient texts, killed him. All that remained was for him to marry his mother, Jocasta, and, wouldn't you know it, by a bizarre combination of fluke circumstances this duly came to pass.

So far so bad. But it gets worse—Oedipus and his mother eventually found out the awful truth. Jocasta hanged herself, and Oedipus, not bearing to look on the unhappiness he had caused, took pins out of her clothes and gouged his eyes out with them. Oedipus must have wished he had lived in a more scientific age, though according to the logic of the story he wouldn't have been able to escape his tragic *moira* even then. In a sense, the fact that he tried to skip his destiny marked him as a "modern man" before his time, with heretical delusions of self-determination. Sophocles thought Oedipus was wrong to try to resist what the deities had decreed, though by their rules neither rebellion nor submission would have affected its occurrence. The uncompromising and inhuman insistence of the ancients that chance utterances of oracles and coincidental alignments of *omens* could define events yet to happen damned the fictional Oedipus as a loser. Not even time travel to a skeptical future could have prevented him from murdering his father and marrying his mother.

His story had been around a long time even when Sophocles got hold of it. Interestingly, one of the reasons the writer may have chosen it was to reassert the values and worldview represented by the old gods in the face of new modes of thinking that were being developed in some of the Greek city democracies. Here and there groups of philosophers were rejecting old superstition-based beliefs and beginning to assert rational theories of existence based on empirical evidence—formulating, had they known it, the basis of the modern scientific method.

It was to be a long time before science would take over as the principal means of examining natural phenomena. Until it did history was to provide numerous examples of how dangerous superstition could be as a means of interpreting the world. Perhaps the most tragic is the Aztec Empire of Central America. In 1519, when the ambitious Spanish adventurer Hernando Cortés landed near modern Veracruz and burned his ships, effectively cutting off his tiny army's escape route, the Aztecs controlled a civilization of millions of souls that was extraordinarily sophisticated by the standards of its day. In some obvious ways it lagged behind Europe, yet it had advanced mathematics, astronomy, and agriculture, it controlled five hundred vassal states, had an efficient social organization and raised cities that rivalled in size, architecture, and organization anything Europe had to offer. Yet this empire was destroyed by around five hundred Spanish soldiers, its population murdered and enslaved, its monuments and culture trashed, because of its failure to see a coincidence for a coincidence.

Cortés didn't know his luck when he set out for Mexico, but the year happened to coincide with a period in the Aztec calendar when the god Quetzalcoatl was predicted to return over the sea from exile in the west. Quetzalcoatl was a serpent god who took other forms. He was depicted in ancient paintings of his return as fair skinned, with what looked like a beard—much the same form, in fact, as Cortés and his soldiers.

Like all the saddest stories, the Aztec downfall had been signaled some years before, by a comet. Moctezuma Xocoyotl, the last emperor of the Aztecs and a former priest, watched the comet with foreboding from his palace roof. It was the first of many other omens of impending disaster in the years following: temple fires, floods, lightning strikes, and the rumored sighting of many-headed people walking the streets.

The Aztecs had built their empire in barely one hundred years. It's a wonder it lasted so long, for Aztec society was predicated on particularly neurotic forms of superstition. Its large priestly bureaucracy (there were five thousand priests in the capital, Tenochtitlan, alone) was so powerful it dominated all aspects of life. Religious intolerance was not an unusual phenomenon in the sixteenth century; but few gods were as vicious and jealous as the Aztec gods. Chief among them was Huitzilopochtli, God of War. In return for providing the world with sunlight, Huitzilopochtli required the fuel of human blood. Each year, in order to pay for their daylight, which in other cultures comes free, the Aztecs had to dash the still beating hearts (the gods' preferred serving suggestion) of many thousands of men onto Huitzilopochtli's altar. As the Aztec Empire became larger, Huitzilopochtli, whose desires were interpreted by the paranoid and neurotic priestly elite, demanded more and more human sacrifices, thousands at a time. The victims were mostly prisoners taken in war, but as there were never enough of these, the Aztecs stage-managed special wars with their vassal states that were nothing at all to do with territory or counterinsurgency or killing enemies. They were called "Flowery Wars" and were arranged like football matches; their purpose—to obtain large numbers of prisoners for use as sacrificial fodder to nourish the Aztecs' gods. As well as the obvious human cost, this practice sowed widespread resentment throughout the empire.

Cortés was cruel and brave and a brilliant opportunist, ever alert to his enemy's weaknesses, but he was also so reckless it's a wonder he and his men didn't die within a few weeks of landing on Aztec soil. Few military expeditions have been embarked upon with such inferior manpower and resources, at so remote a distance from sources of supply, and with so little tactical logic. Historians have made much of the debilitating effect on the Aztec fighting men of Spanish cavalry, Toledo steel swords and firearms, but in fact the Spaniards possessed only fifteen horses and early-sixteenth-century firearms were notoriously unreliable. However deadly that Toledo steel, the Aztecs were fearless and fierce, and were fighting on their home turf. Most importantly, they were numerous. At any point, had they chosen to decisively engage their enemy, they could have triumphed by sheer weight of numbers. But the Aztecs didn't fight in any concerted way until it was too late. Cortés's most important weapon turned out to be luck: the luck of the Devil.

While Cortés pushed resolutely ahead, Moctezuma was paralyzed with indecision, terrified by the possibility that the Spaniard was an incarnation of the returning deity Quetzalcoatl. Cortés, quick to realize how the coincidence could be exploited to his advantage, negotiated alliances with the tribes that had been so cruelly exploited by the Aztecs—the Tlaxcalans, Cempoalans, and others—in exchange for promises he later failed to honor. Then his ragtag army marched on the Aztec lake city capital Tenochtitlan, a city that, with a population of three hundred thousand, was larger than any in Europe at the time. When he arrived, Moctezuma, still indecisive, invited him and his soldiers into the city as his guests, and later into the royal palace itself (Moctezuma's palace guard alone was larger in numbers than the entire Spanish expedition), where the ungrateful Spaniards arrested him.

There were other factors, including a devastating smallpox epidemic brought by the Europeans, yet there's no escaping the conclusion that Moctezuma and his empire were brought down by tricks of the imagination; by superstition and the value it attributes to random signs, portents, and things that go bump in space.

On the face of it modern society would appear to have moved

away from such destructive superstition. Rational scientific analysis is the modern means of measuring and interpreting cosmic debris. In fact in 1996 a newly discovered asteroid was named Skepticus in honor of the Committee for the Scientific Investigation of Claims of the Paranormal (which publishes the magazine *The Skeptical Inquirer*) and a second rock after CSICOP's founder Paul Kurtz.

Yet only a year after that ceremony, when the comet Hale-Bopp appeared in the night skies in 1997, thirty-nine members of a religious mission called Heaven's Gate cleaned their communal house at Rancho Santa Fe, California, donned identical black running shoes and tracksuits bearing comet swoosh badges and "AWAY TEAM" patches, attached identification to themselves, dosed each other with a cocktail of apple sauce, vodka, and phenobarbital, placed plastic bags over their heads and lay patiently on their beds to die. In videos they made before their suicide they recorded their belief that the comet was the marker they had awaited for so long and that they were ready to shed their "containers" and leave the planet in a space-craft sent by beings in the "level above humans." No use telling them it was all just coincidence.

Commenting on the tragedy, University of Southern California professor of religion Robert S. Ellwood said, "These people come from a nineties kind of culture, with all its hardware and world views, but they have hewed to the traditional apocalyptic scenario: that radical changes are imminent and foretold by signs in the heavens."

In a society that reckons itself to be founded on rational scientific notions, 17 percent of Americans claim they have seen a ghost, 10 percent say they have communicated with the devil, and four million claim to have been abducted by aliens. Evidence that the paranormal is not only alive and well, but doing big business, can be easily found in the astrology columns in every magazine, in the psychic consultation ads in newspapers, in the popularity of creationism (which holds that the Earth was made in seven days), in the hiring by big business of consultants in dowsing and feng shui.... Skeptical Americans were shocked in 1986 when a Philadelphia jury awarded more than $900,000 in damages to a woman who claimed her psychic powers had been damaged during a CAT scan at a University Medical School. Her complaint was supported by the "expert" testimony of a doctor.

Our obstinate attraction to paranormal explanations is explained by psychologist Susan Blackmore as a natural tendency to try to understand the world by making connections between things such as dreams and events, or star formations and our love lives. Coincidences are ready-made connections; all that remains is for us to label them prophetic.

Psychologists call the attempt to link random events with our own thought processes the "illusion of control." Dr. Blackmore gives a simple example that we have all experienced—willing traffic lights to change as we approach them. If the lights do change we get a pleasant lift, but people often report such coincidental events as evidence of psychokinesis, or mind over matter. In other words, physical objects have somehow reordered themselves in accordance with a thought in someone's mind. Investigations have revealed that people who report such powers routinely ignore those occasions when the lights do not change, if they notice them at all.

"We like to think we can control the world around us by observing coincidences between our own actions and the things that happen," says Dr. Blackmore. "Belief in psychic events may be an illusion of causality."

Psychics respond to such criticisms by asserting that human intuition is a greater force than science allows, yet altogether too subtle and idiosyncratic to be subjected to the empirical tests that science demands. "We are more powerful than we know," says Craig Hamilton Parker, who describes himself as a psychic. "In my work as a medium I have found the spiritual state of people influences the world around them. We can change events with the power of thought. Mind can influence matter. With training we can make our own world and the whole world better."

He says there is no such thing as chance, merely human will.

"Coincidences are nothing of the sort, just our awareness that the outer world is actually an inner world. Coincidences synchronize the inner and the outer worlds."

Is this correct? Science surely has an answer to that question.

. 4 .

IT'S A SMALL
UNIVERSE—
COINCIDENCE AND
SCIENCE

Science writer Arthur Koestler called coincidences "puns of destiny." The Nobel Prize–winning physicist Wolfgang Pauli thought they were "the visible traces of untraceable principles." Both men believed a mysterious and seemingly magical force was at work in the universe imposing order on the chaos of human life. This powerfully evocative idea violently jars with the skepticism of classical science. If there is any truth in the idea, which emerged with a vengeance in the philosophical/scientific debate of the 1950s ("the paranormal equivalent of a nuclear explosion," said one reviewer of Pauli and of Carl Jung's treatise, *Synchronicity: An Acausal Connecting Principle*), then all kinds of phenomena previously damned as mere coincidence, such as telepathy and precognition, represent themselves for serious reassessment. To many rational minds the idea of meaningful coincidence is still out there with the divining rods and the ectoplasm cowboys and not a subject for serious research; nevertheless the idea that there might be a link between psyche and matter has attracted some pretty powerful minds.

There is no shortage of anecdotal accounts of people with unusually powerful telepathic abilities. The anthropologist Laurens van der Post claimed that the Bushmen of the Kalahari knew when a fellow hunter had made a kill fifty miles away. He said these hunters took it for granted that their families waiting for them at the village would know if and when they were coming back with a deer.

How do we assess such a story? It hasn't been rigorously tested, certainly, but is it true? It could be a delusion on the part of the Bushmen, or a false notion on the part of the reporter. Rupert Sheldrake, who employs the Bushmen story to illustrate his theories about telepathy, thinks it is reliable. "These are survival skills," he says, "and they were probably common to all humans in primitive societies." As a scientist with a Cambridge University Ph.D. in biochemistry, Sheldrake is unusual in these opinions. He makes a point of studying subjects such as telepathy, which other scientists dismiss as coincidence, saying he finds them more challenging. "Though [intuitive abilities] have ceased to be important to people in modern society they remain in most of us in vestigial form." He says that the least intuitive people, ironically, are white male academics. "Mothers and children are intuitive, and businessmen, who work with partially known factors and uncertainty—it's an essential part of capitalism."

His experiments to test the possibility of human telepathy are controversial in the scientific community. Sheldrake has endeavoured to measure whether we sometimes know who is calling as the phone rings, and in his book, *The Sense of Being Stared At,* he examines the possibility of people being able to sense when someone is watching them behind their backs. He suggests that incidents of intuition such as this might have an invisible causal influence. "Somehow our intentions, and our attention, reach out to touch what we are looking at."

Words such as "telepathy" are taboo in classical science, yet Sheldrake points out that similar invisible influences—radio waves, for example—are well known to it. With the exception of light, the average person has only a dim perception of most of the elements in the electromagnetic spectrum, yet scientists routinely detect X rays, gamma rays, radio and microwaves and read them for the information they communicate about astral events in distant space.

On a more mundane level Sheldrake has been examining apparent precognitive powers in animals. How do animals seem to know when their owner is about to arrive home? He claims his experiments show there may be a telepathic element in this. In China, Sheldrake says, it is widely accepted that animals behave nervously prior to earthquakes. Chinese seismologists have asked members of the public to report the unusual behavior of rats, fish, birds, dogs, and horses. As a consequence, says Sheldrake, they are the only seismologists in the world to have accurately predicted earthquakes (though they failed to predict the 1976 Tangshan earthquake in northern China, which killed almost a quarter of a million people). "No one knows how animals do it," Sheldrake says, "whether it's tremors, gasses, or something more mysterious like precognition. It's not just coincidence."

Sheldrake thinks some coincidences can be explained by his theory of morphic resonance, which postulates telepathic-type connections between organisms, and fields of collective memory within species. Ideas are simply "in the air" and if you are tuned to the right station you will pick them up.

To illustrate how this works among animal species, Sheldrake quotes the example of a group of laboratory mice in London that were taught to improve their maze-running skills. Almost immediately, untutored mice in a similar maze in a Paris lab were reported to have achieved the same navigational feat. Another example concerns a monkey on a South Pacific island that discovered potatoes tasted better if it washed them in the sea before eating them. Shortly after this behavior was observed, it was reported that monkeys throughout the archipelago had begun washing potatoes, too.

"Descartes believed the only kind of mind was the conscious mind," says Sheldrake. "Then Freud reinvented the unconscious. Then Jung said it's not just a personal unconscious but a collective unconscious. Morphic resonance shows us that our very souls are connected with those of others and bound up with the world around us."

Sheldrake has been criticized by mainstream scientists for his New Agey fixations. Robert Todd Carroll, a professor of philosophy who edits the Skeptic's Dictionary Web site, calls Sheldrake a metaphysician rather than a scientist, casts doubt on the rigor of his experiments and accuses him of confirmation bias (the tendency to report only evidence that coincides with the researcher's pet theory). The South Pacific monkey story, says Carroll, is anecdotal and unreliable.

Sheldrake does not walk his astral plane alone, however. Nor does he walk where no scientist has gone before. In fact, the path is already well trodden by mavericks like the Austrian biologist Paul Kammerer and the psychologist Carl Jung, who suspected there might be more to apparently coincidental events than meets the eye and were prepared to put their reputations on the line by publishing their theories.

In the early 1900s Paul Kammerer kept journals in which he faithfully recorded every coincidence he experienced, from the incredible to the downright mundane. Kammerer was interested in the fact that coincidences tended to cluster in groups. In 1919 he introduced *The Law of Seriality*, in which he conjectured that these clusters were evidence of some deeper force at work that we do not see. Coincidence clusters were like ripples on the surface of a pond, the only observable evidence of a general connecting principle of nature, a major force in the universe similar to gravity. But whereas gravity works only on objects with mass, seriality affects both objects and consciousness, bringing things together by affinity. Kammerer thought the peaks that we call coincidences are glimpses of a hyperconnected universe whose weblike workings we are only vaguely aware of and nowhere near understanding. "Seriality is ubiquitous in life, nature, and cosmos," he said. "It is the umbilical cord that connects thought, feeling, science, and art with the womb of the universe that gave birth to them."

He concluded, "We thus arrive at the image of a world mosaic or cosmic kaleidoscope, which, in spite of constant shufflings and re-arrangements, also takes care of bringing like and like together."

Kammerer lived at a time when the classical laws of physics were starting to groan under the strain of startling new discoveries and ideas. The clockwork universe had been ticking along reliably since the seventeenth century, when René Descartes, Thomas Hobbes, Isaac Newton, and others established its rational basis in human thought. In the nineteenth century, matter was held to be the fundamental and final reality. Scientists saw the universe as a grand machine governed by immutable laws, every part interacting with every other part in a logical and predictable fashion. Time ticked reliably along from past to present—you could set your watch by it. Effect followed cause in reassuringly strict sequence. You could ascertain the cause of something by examining the effect, and the laws affecting one part of the machine applied to all parts of the machine. It was a reductionist approach: you could analyze anything by breaking it down and examining its parts.

The wrench in the works was human consciousness, which stubbornly refused to be broken down. Where did sentience, self-awareness, and free will fit in to a purely material universe? Just how the mind works and what thinking is are two of the profoundest mysteries. The attempt by classical science to explain the human mind away as a sort of fancy computer, quite apart from being intrinsically unattractive, was unconvincing.

The twentieth century brought with it new ways of looking outward into space and inward into the atom. Both directions offered astonishing revelations that confounded classical realities. We learned that energy and matter were two different expressions of the same thing ("a somewhat unfamiliar concept for the average mind," said Einstein with understatement), that light was deflected by gravity, and that time, which previously had waited for no man, was prepared to make an exception if he was traveling at the speed of light. Light itself was revealed as contrary, behaving sometimes like a wave and sometimes like a stream of particles, depending on how it was observed. Out in deep space unthinkably dense black holes gyred and roiled, sucking in stars and light, distorting space and time around their circumferences and emitting the deepest roar in the Universe.

Inside the atom, formerly thought to be an indivisible ball (hence the name, from the Greek *atomos*, meaning uncuttable), there was revealed a miniature universe in which things happened that contradicted the classical laws pertaining to the big world. Here gravity held no sway because atoms were held together by their own, vastly stronger special forces, cause and effect didn't seem to apply, and the exact states of particles could never be predicted. The behavior of a light photon encountering a sunglass lens is impossible to predict. We know the probability of photons bouncing off the surface and also the probability of them going straight through, but it is impossible to predict what any individual photon will do or know why it has chosen that behavior. Science, with its dependency on hard measurable facts, found itself treading water in a probabilistic universe that confounded the old certainties.

Electrons, those tiny particles that exist in orbits around an atom's nucleus, exhibited the same wave/particle duality as light, suggesting that in a microscopic sense, all matter is wavelike. Electrons were very mysterious; Einstein called them "spooky." They appeared able to exist in twenty places at once (quantum superposition), they would suddenly change their behavior for no causal reason, and if a pair of linked particles were separated they exactly mirrored each other thereafter (quantum entanglement), whether they were two feet or a billion miles apart. An experiment that changed the state of one would be instantly reflected by a corresponding change in the state of the other, the information having passed between them across any distance instantaneously. Each particle seemed to "know" what the other was doing. The phenomenon is very difficult to explain as it violates Einstein's law that nothing can travel faster than the speed of light. Scientists have used the word "telepathy" to describe it and have even speculated that the particles' separation may be an illusion.

More alarming for traditional scientists was how personal the study of the atoms' interior parts was becoming. As soon as a subatomic particle such as an electron was measured (i.e., observed) it changed its behavior. If you tried to measure a particle you found something that looked like a particle, otherwise it behaved as a wave. Things changed when you looked at them so you could never know what they looked like before you looked. Interpretation was necessary. Scientists were forced to be subjective—that intimate adjective that also defines the essence of consciousness and coincidence. Quantum physics seemed to be teaching us that at the microscopic level there may be no objective reality; that what we observe is always affected by the presence of the observer. Wolfgang Pauli, the Nobel Prize–winning physicist who first postulated the existence of the neutrino in 1931, said: "On the atomic level the objective world ceases to exist."

With science getting weirder and weirder all the time, the claims of ESP (extrasensory perception) and psychokinesis, collectively known as psi, are starting to sound rather tame. It's almost as though science has decided to take on psi at its own game of fantastic unbelievability and is beating it hands down.

Look at the Narnia-like world of the atom; it's a place so small you can't even see it, a world that from our remote distance seems utterly condensed and claustrophobic, yet the closer you get to its paradoxical and common-sense-defying reality, the wider its wide open spaces are revealed to be. The atom is about .000004 of an inch across, but 99.99 percent of its volume is empty space. If we drew an atom to scale, making its nucleus .3937 of an inch then its electrons would measure less than the diameter of a hair, and the entire atom's diameter would be greater than thirty-three football fields laid end to end. In between—nothing. Scientists believe that in a human body the relationship between so-called mass and space is 200 trillion to 1. Einstein calculated that if the space between all the atoms in all the human beings on Earth were removed, leaving only concentrated matter, you would be left with something about the size of a baseball (though a lot heavier).

If a neutrino, one of Pauli's tiny, chargeless, and virtually mass-less particles that are created by nuclear explosions on distant stars and blow through space in their billions, were able to see as it hurtled toward Earth at the speed of light, it would register our planet only as a patch of barely differentiated haze, through which it would pass like a bullet, not interacting with it at all.

So if this bedrock we believe we stand upon is little more than an illusion, what's left? Energy is left—lots and lots of it. That's something we all know is abundantly packed inside every atom. Physicist Max Planck said, "Energy is the origin of all matter. Reality, true existence, is not matter, which is visible and perishable, but the invisible, immortal energy—that is truth."

We are made of atoms, which are made up of tiny packets of electromagnetic force, all of them interrelating and communicating with each other in highly complex ways. These charged elementary particles can transform into each other and carry all the information necessary to explain all of existence. Our bodies are made of the same stuff as Mount Everest and the Pacific Ocean. If you look at us on an atomic scale then we and the universe comprise a seamlessly integrated web; it's all energy and information swapping back and forth. As the astronomer James Jeans put it, "The universe looks less and less like a great machine and more and more like a great thought."

The question is: whose thought? Aliens? Leonard Nimoy? Albert Einstein said, "After years of thought, study and contemplation, I have come to the conclusion that there is only one thing in the universe and that is energy—beyond that is a Supreme Intelligence." It should be pointed out that Einstein's supreme intelligence, whom on other occasions he hasn't been shy to call "God," was nothing like an angel-and-trumpet deity, but something more akin to a perfectly crafted physical law. According to the *Wall Street Journal*, however, modern science is sufficiently tolerant of transcendental ideas for 40 percent of American physicists, biologists, and mathematicians to declare without embarrassment their belief in God.

Spirituality is an important trigger in coincidence experiences, because it is exactly that kind of subjective response that brings converging events to meaningful life. The German philosopher Arthur Schopenhauer saw coincidences as a reflection of the "wonderful preestablished harmony" of the universe. Writing in 1850, he expressed the idea that we were not just motivated by physical causality. He said coincidences constituted a "subjective connection" to the environment. They were important because they were tailor-made to fit individuals, and only relevant to those who experienced them.

So there is nothing new about the idea that all things in the universe have some kind of correspondence and sympathy one with the other. In fact Hippocrates got there way before Schopenhauer, in the 5th century B.C. He believed hidden affinities held the universe together. "There is one common flow," he said, "one common breathing, all things are in sympathy. The whole organism and each one of its parts are working in conjunction for the same purpose . . . the great principle extends to the extremest part, and from the extremest part it returns to the great principle, to the one nature, being and not-being." Or, as the astronomer Carl Sagan put it, "In order to make an apple pie from scratch, first you must invent the universe."

The Swiss psychologist Carl Jung was influenced by Schopenhauer and Kammerer, and also by Eastern religions and philosophies, which have similar ideas about the universal interconnectedness of things, and which see the material world as *maya*, an illusion. True contentment in life can only come by shedding the prison of the ego and surrendering unconditionally to the great flow. For many years Jung had been intrigued by the coincidences related to him by his patients, though the word "coincidence" seemed increasingly inappropriate as many of them were "connected so meaningfully that their 'chance' concurrence would represent a degree of improbability that would have to be expressed by an astronomical figure." Like Kammerer and Schopenhauer, he too saw them as a reflection of universal connectedness: "The universal principle is found even in the smallest particle, which therefore corresponds to the whole."

Jung was discontented with what he called the "Godless, meaningless, clockwork universe of modern science," though a series of dinners with Albert Einstein, at which the great scientist revealed the latest insights into the wonderful, mysterious realms of relativity and quantum mechanics, inspired him to devise a philosophical framework that could explain the significance of coincidences and the force that generated them in the first place. Quantum mechanics for Jung was proof that at a fundamental level the universe didn't behave like a machine at all. Jung didn't want to dethrone classical science; just show there might be more to it. He believed, too, that science and spirituality should walk hand in hand, a belief shared by Einstein.

One of Jung's most useful legacies is the word "synchronicity," which goes beyond the strict meaning of coincidence to include our subjective human experience of chance events. Synchronicity refers to coincidences that are meaningful to the percipient, in which something other than the probability of chance is involved. This meaningfulness can only be judged subjectively and is therefore open to interpretation—a constraint analogous to that of the modern subatomic physicist pondering whether a particle is a particle or really a wave, and what he might have done to change it from one to the other.

In 1952 Jung teamed up with another brilliant visionary, physicist Wolfgang Pauli, to publish *Synchronicity: An Acausal Connecting Principle*. Jung defined synchronicity as: "The coincidence in time of two or more causally unrelated events which have the same meaning." The Jung/Pauli relationship was in itself like a synchronistic correlation: two unrelated spirits from ostensibly incompatible disciplines, one a philosopher/psychologist, the other a quantum physicist, together finding a deeper meaning than either of their disciplines allowed, a new reality they called the *unus mundus* (unified world), in which mind and matter were united. Both men experienced vivid dreams.

Jung was one of the first modern thinkers to take dreams and symbols seriously. He introduced the idea of the collective unconscious—the distilled memory of the human species from its primitive beginnings to the present day—to which he believed we all have intuitive access and which has important transforming effects on us in times of dramatic import in our lives. It communicates with us through dreams, visions, and meaningful coincidences, which may range from a truly unexpected correlation to a simple throw of the I Ching yarrow stalks. Jung was a rare thing in the modern age—a philosopher who allowed for and even explained the paranormal.

The collective unconscious is a psychological substratum built into the inherited brain structure, consisting of cultural metaphors common to all humanity, expressed in stories, myths, symbols, and ideas. Jung called them archetypes. They are not things we can consciously understand; rather they are manifestations of psychic energy. We don't necessarily think about archetypes, yet they are immensely evocative themes lurking subliminally in our minds. For example, water is a metaphor for life; fighting a dragon is a struggle between good and evil. Other examples among many archetypes are the mother, the hero, the maiden, the trickster and the hermaphrodite. Jung believed we all have access to this common source of resonant ideas, in the same way that subatomic particles share their bundles of energy and information—think of it as a sort of cosmic computer.

Our access to these common archetypes has nothing to do with conscious control. At times we may even fear them. Given the paramount importance modern society places on rational self-control, Jung said, we have a tendency to repress archetypes and deny their existence. Despite this, in certain circumstances they will synchronistically manifest themselves in both matter and mind simultaneously. When this happens it delivers to us a sense of numinousness, or deep spiritual significance, often overwhelming, of participating in one of Jung's "acts of creation in time"; a sense of absolute cosmic authority.

The most famous example of this archetypal synchronicity is the scarab beetle that appeared at the window of Jung's study during a consultation. This must be the least private consultation ever held, since it has been repeated so many times. Nevertheless it serves to illuminate a complicated idea.

The patient was a woman who up until that point had refused to believe that anything could help her condition, which was complex and refractory. Jung was the third doctor she had seen and up to this point no progress had been made. "Evidently something quite irrational was needed that was beyond my powers to produce," he said.

The woman was in the middle of recounting a dream to Jung in which she had been given a golden scarab, when a tapping at the window distracted them both. Jung opened the window and in flew a scarab beetle, the local version of the insect in the patient's dream. In Ancient Egypt the scarab was a symbol of rebirth. "Contrary to its usual habits," said Jung, "it had evidently felt an urge to get into a dark room at this particular moment."

This symbolic event shocked the patient into the realization that she could control her condition. "Now she understood how all kinds of connections might exist and how they would explain a great many things if they did. She recovered quickly."

In his early work Jung thought archetypes were exclusive to the human mind. Later he suggested that they shaped matter as well as mind. In other words, archetypes were elemental forces that played a vital role in the creation of both the world and the human mind. Synchronicities were events in which the inner and outer worlds, the subjective and the objective, the psychic and the physical, briefly united.

Jung wrote, "We delude ourselves with the thought that we know much more about matter than about a 'metaphysical' mind or spirit, and so we overestimate material causation and believe that it alone affords us a true explanation of life. But matter is just as inscrutable as mind."

You don't have to believe it, of course, and many don't. Jung was a psychologist and psychologists have a long tradition of being criticized by those in the more rarified zones of science for their metaphysical theorizing and predilection for unprovable anecdotes. Such weird notions as thought patterns affecting matter are out of bounds in classical science.

But classical science is finding its own presumptions and methods under attack today. This is Dean Radin, director of the University of Nevada Consciousness Research Laboratory, in his book, *The Conscious Universe: The Scientific Truth of Psychic Phenomena*: "When modern science began about three hundred years ago, one of the consequences of separating mind and matter was that science slowly lost its mind."

Radin has muscle to back his cockiness. He's one of a new and growing breed of researchers out to prove the validity of psi, who show none of the amateurishness previously associated with enthusiasts of the paranormal. Radin is more than comfortable with the rigorous standards demanded by skeptics. His experiments, when applicable, incorporate control group studies, double-blind controls, and random testing; in other words, they honor scientific standards designed to reduce error, self-deception, and bias. And he provides full public access to his methods and results in order that they may be reviewed and retested by peers, a tried method of weeding out bad science. In fact, the Committee for the Scientific Investigation of Claims of the Paranormal (CSICOP), the publisher of *The Skeptical Inquirer,* is offering skeptics a special revision course, called The Skeptic's Toolbox, to meet the new threat.

Radin has recently been doing research with Professor Dick Bierman, of the University of Amsterdam, into presentiment, which is defined as the apparent psychological effect of a future emotional cause. The results would appear to turn the normal order of cause and effect back to front.

In the experiment, subjects viewed a series of randomly selected images on a screen, some very neutral and some violent or sexual. The subjects' emotional response to the images was measured throughout by means of a skin-conductance-measuring device. The subjects responded more strongly to emotional images than neutral ones, but in the case of the emotional stimuli the reaction started a fraction of a second before the image appeared on the screen. The experiment suggests that people can somehow "see" emotionally charged images before they appear.

When Professor Bierman repeated the experiments with subjects in a brain imager, the emotional responses began up to four seconds before the stimuli.

Meanwhile up in the outer limits of Mad Science Narnia things are getting weirder by the day. Those wild-eyed physicists with the up-all-night hair and the untestable versions of existence derived from preposterous mathematical formulae only wild-eyed scientists with up-all-night hair can understand, no longer sound like the hard-headed pragmatists of mechanistic science. They sound like the Pythia of the Delphic Oracle, spouting oracular riddles for the perplexed hoi polloi to ponder.

Physicist David Bohm suggested that the Universe is a vast hologram in which, just as a hologram cut up contains the entire image in every piece, every part contains the whole order. This "explicate order" was a projection from higher dimensional levels of reality. Like Jung, Bohm believed that life and consciousness were enfolded in every level of matter. He said the separation of matter and spirit was an abstraction. Bohm's holographic paradigm, a popular idea among many scientists today, suggested a universe of infinite interconnectedness.

Then there's hyperstring theory, an imaginative idea that reconciles incongruities between relativity and quantum theory at the expense of adding six dimensions to our current four, some of them microscopic and curled up on themselves. It's a universe in which the notions of space and time disappear and energy is represented as tiny strands of spectral string, which crackle and fidget realistically in computer simulations but in reality cannot be seen.

Are ten dimensions enough? Some scientists have suggested there may be sixteen or seventeen. Scientific philosopher David Lewis thinks there could be an infinite number of them. Others say there's only one, and that this one is infinite enough for all of us—and our theories. Measures of the cosmic background radiation (the echo of the Big Bang) indicate that it is so large that all possible arrangements of matter must exist within it. In fact this universe contains, in a galaxy somewhere around $10^{10^{28}}$ light-years away, an exact replica of our own planet and everything in it. Surely that's the coincidence to end all coincidences!

There may be logic in these theories, but is there reason? And if the reason that is there applies in the microscopic world of particles or the macroscopic world of galaxies, how does it apply to us who are stuck in the medium-size world of lawn mowers and armchairs? In this medium world we have to keep our feet on the ground.

Our advice is keep your head and don't accept any rides from aliens. Keep looking out for coincidences, though, because it's healthy—or so says Professor Chris French. "We have been successful as a species precisely because we are good at making connections between events," he says. "The price we have paid is a tendency to sometimes detect connections and patterns that are not really there."

So caution is advisable, and beware that other human weakness defined by psychologists: apophenia, the spontaneous perception of connectedness and meaningfulness of unrelated phenomena. People with mental disorders seem particularly susceptible to it apparently. In fact there's a fierce debate blowing currently about whether these experiences are a symptom of mental illness, or a cause.

In his book, *The Challenge of Chance,* science writer Arthur Koestler said that at the very least coincidences "serve as pointers toward a single major mystery—the spontaneous emergence of order out of randomness, and the philosophical challenge implied in that concept. And if that sounds too rational or too occult, collecting coincidences still remains an amusing parlor game."

.5.

COINCIDENCE IN THE DOCK

In March 1951 two boys called Dennis were born; one in California, one in Dundee, Scotland. Both were naughty boys, both favored striped jerseys. Both were named Dennis the Menace by their artist creators. Fifty years on, the American Dennis, by Hank Ketcham, is still a popular newspaper strip, and the Scottish Dennis still features every Thursday in D.C. Thomson's *The Beano*. Both creators identified the similarities as a coincidence and agreed not to encroach on each other's market.

Not all similarities between products are resolved so sweetly. Not all similarities between products are coincidences. . . .

And what do you do if someone accuses you of stealing their brilliant original idea for a cartoon character, or magazine, or catchy funeral march. How do you prove that you are the rightful owner of a brain wave, that any similarity between your idea and someone else's is pure coincidence? It's a legal minefield that has blown the legs off many a litigant.

Musician Mike Batt had to pay a small fortune to settle a bizarre dispute over who owned the copyright to silence. Batt was accused of plagiarism by the publishers of the late U.S. composer John Cage. Batt's alleged crime was to place an entirely silent track on his 2002 album, *Classical Graffiti*. He called the track "A One Minute Silence" and credited it to Batt/Cage.

Cage had written his own silent composition, "4'33"," back in 1952. At its first performance, by pianist David Tudor, at Woodstock in New York, many in the audience reported that they had failed to hear the work. Cage's composition had three movements of differing lengths. The total running time was "4'33"."

Batt attempted to prove his silent track differed from Cage's by staging a performance of the piece by an eight-piece ensemble. He said: "Mine is a much better silent piece. I have been able to say in one minute what Cage could only say in four minutes and thirty-three seconds." Cage's publishers responded by hiring a clarinettist to perform Cage's silent composition.

In the end Batt lost the legal battle. He did succeed in proving that silence is golden—but only for Cage's publishers, to whom he handed over a six-figure sum in an out-of-court settlement. He subsequently released "A One Minute Silence" as a single. It was never heard on *Top of the Pops*.

Disputes about the ownership of silent compositions are rare. When the music becomes audible, so does the rustle of lawsuits.

George Harrison claimed that the similarity between his number-one hit, "My Sweet Lord," and the Motown classic, "He's So Fine," was mere coincidence. The judge disagreed, saying it was "perfectly obvious the two songs are virtually identical." The judge accepted that Harrison had not consciously set out to appropriate the melody of "He's So Fine" for his own use, but said that that was not a defence.

Harrison had admitted that he had heard the Motown song prior to writing "My Sweet Lord," and that therefore his subconscious mind knew the combination of sounds. The judge decided that Harrison was guilty of "subconscious plagiarism." He said it was not an area in which precise measurement could be made, but concluded that three-quarters of the success of "My Sweet Lord" was due to the plagiarized tune, and one-quarter of that success was due to Harrison's name and the new words he had written. He concluded that $1,599,987 of the earnings of "My Sweet Lord" were reasonably attributable to the music of "He's So Fine."

Consciously or subconsciously, deliberately or coincidentally, songwriters seem to make a habit of imitating each other's efforts.

If you were to sit at your local church organ, pull out all the stops and play the opening few notes of Bach's Fugue in E Flat, while simultaneously humming the familiar tune to the hymn "O God Our Help in Ages Past," you would not scare the congregation. The opening notes are identical. Just coincidence?

Try another one. Dust down an old 78 rpm copy of the American song "Aura Lee," written by George R. Poulton and popular during the Civil War. The tune bears an uncanny resemblance to the slightly more recent "Love Me Tender" credited to Elvis Presley and Vera Matson. Were all the writers involved tapping into the same universal creative consciousness? Was it just coincidence?

Similar hard-to-answer questions have been raised in relation to the highly successful musical-writing career of Sir Andrew Lloyd Webber. Cabaret act Kit and the Widow entered the territory with their withering parody of the multimillionaire's oeuvre entitled "Somebody Else." The song works its way through some of Lloyd Webber's most familiar and successful show tunes, pointing out some startling similarities—such as those between:

- "Memories" from *Cats* and Ravel's *Bolero*
- "Jesus Christ Superstar" and a tune by Bach
- "I Don't Know How to Love Him" from *Jesus Christ Superstar* and a Mendelssohn violin tune
- "Oh What a Circus" from *Evita* and Bach's Prelude in C

These are just a few of the coincidences between Lloyd Webber tunes and the works of the great composers turned up by Kit and the

Widow's research. Did Kit and the Widow think he had received musical messages in his sleep from the great masters? Or, given that there are only seven different white notes on the piano and a smattering of black notes, were the similarities the result of pure serendipity—the synchronistic workings of great musical minds? They were not in fact persuaded by either explanation. They prefer to think a little bit of artful borrowing has been going on, made possible by the fact that the copyright on most of the great classical canon has long since expired.

Kit and the Widow say Lloyd Webber has admitted to them that some of the similarities between his songs and previous works are "too close for comfort." But he also teasingly pointed out that their parody had "missed some of the best ones."

Issues of coincidence versus plagiarism emerge almost as frequently in literature as they do in the world of music. V. S. Naipaul, winner of the Nobel Prize for Literature, famously declared that the novel was dead—that all available plot possibilities had been thoroughly exhausted. Not that that prevented him writing one further novel himself.

If we assume that the novel, having been briefly resurrected to allow Naipaul a last hurrah, is once more deceased, it is hardly surprising that we find familiar stories popping up from time to time in newly published works of fiction. Not all writers can claim their every publication is a work of truly original genius. Even Jeffrey Archer has been accused of plagiarism. And in this he is in the company of no lesser writer than William Shakespeare.

Writers throughout the ages have faced accusations that they have done a little "borrowing" from the great lending library of other writers' ideas. Shortly after his death in 2002, the Spanish Nobel laureate Camilo Jose Cela was accused of being both a cheat and a plagiarist. It was said that Cela regularly used ghostwriters for most of his career, including for his *The Cross of Saint Andrew,* which won him Spain's prestigious $450,000 Planeta Prize.

It is alleged that in *The Cross* and other books, ghostwriters supplied the plots and characters, which Cela incorporated into his own prose. "Cela was a great prose writer with an exquisite style but plots and arguments were not his strong point," said his accuser, journalist Thomas Garcia Yebra.

In the case of his Planeta Prize–winning novel, it is further alleged that the ghostwriter had plagiarized the unpublished manuscript of a schoolteacher that had been submitted for the same literary competition. Her claim has been rejected in the courts although appeal judges found "innumerable coincidences" between the two works.

Cela once said that he would like his epitaph to read, "Here lies someone who tried to screw his fellow man as little as possible."

British novelist Susan Hill feels she has been screwed, but only by a series of unfortunate coincidences. The chain of events began in 1971 when she published her novel, *Strange Meeting*, about two young soldiers in the trenches during the First World War.

"In that ultrasensitive state immediately following the completion and publication of a novel, I was plunged into depression when another, about the love of two young soldiers in the trenches of Flanders, Jennifer Johnston's *How Many Miles to Babylon?* came out shortly after mine."

Years later she had another idea for a story—which became a novel she called *Air and Angels*, which she finished and sent to the publisher in May 1990. It was set in Cambridge around 1912. One of the central characters is a don and cleric who falls in love with a sixteen-year-old girl.

"One fine Sunday morning we were having coffee at a café table overlooking the Royal Shakespeare Theatre and the River Avon at Stratford . . . when my husband looked up from his paper and said quietly, 'There is an interview with Penelope Fitzgerald here that you had better read.' Alerted, though somewhat puzzled by the seriousness of his tone, I set aside my own paper and did so. I discovered that Mrs. Fitzgerald was about to publish a new novel called *The Gate of Angels*. Its hero was a clergyman with a scientific bent who falls passionately in love with a very young girl. Its setting, Cambridge, circa 1912."

Hill and Fitzgerald had met once but never spoken or corresponded about their work. Both novels were published and, coincidentally, sold well. But bad luck, they say, comes in threes.

Hill had set her heart on writing a novel about Captain Scott and his companions and their journey to the South Pole. Her research completed, she settled down to start writing.

"Just before Christmas, on the 8:50 from Oxford to Paddington, I opened my copy of *The Bookseller* and saw an advertizement for a new novel by Beryl Bainbridge called *The Birthday Boys* based on the last voyage of Scott and his companions to the Antarctic. Two years' work gone down the tube."

In 1988 two books were published whose contents were very different yet whose covers bore remarkable similarities. Both Marianne Wiggins's novel, *John Dollar,* and Tim Robinson's guide to Ireland's Aran Isles, *Stones of Aran,* had covers that featured a blue dolphin, a black-and-white compass and a map.

The publishers of *John Dollar,* Secker and Warburg, were angry, claiming their cover had been widely distributed within the trade months before. John Caple, the artist who designed the cover for *Stones of Aran,* said he had never seen the other cover.

Fiona Carpenter, art director of Viking, who published the guidebook, said it was just "a very unfortunate coincidence."

Was this simply coincidence, or was it a slightly more subtle example of the technique employed by the junk shop in Clapham that shamelessly called itself Harrods, reproducing the famous colors and typeface? When Harrods threatened legal action it changed its name to Selfridges, claiming the name was valid because it did sell fridges.

When a case of theft of intellectual property comes before a court, the judge or jury must decide if the alleged copying is, in fact, nothing more than coincidence. They must consider what the chances are of someone completely independently coming up with an almost identical design or invention or, indeed, name for a store.

In 1998 director Mehdi Norowzian took brewer Guinness to court claiming a commercial for their famous brew was a copy of his short film "Joy." Norowzian argued that the ad, which showed a man dancing around a pint of Guinness, was a substantial copy of his film and not just "a repetition of an idea." But the judge ruled against Norowzian and ordered him to pay costs to Guinness.

It is probably no coincidence that the people who tend to come out best from litigation over infringement of copyright are the lawyers.

As we've seen, Swiss psychologist Carl Jung had another possible explanation for how two people might come up with the same creative idea—his theory of the collective unconscious that people tap into: "a force of nature which drives us to come to the same conclusions to the same problems, to follow the same creative processes."

Plagiarism even raises its ugly head in the sublime world of laughter. Ownership of jokes, one-liners, and sketch ideas can be aggressively disputed.

Can more than one person come up with exactly the same joke, by coincidence? Kit and the Widow had one of their comic creations apparently stolen from under their noses, passed round the neighborhood and then served back to them cold. Kit Hesketh-Harvey recalls, "The Lloyd Webber musical *Aspects of Love* had controversially cast Roger Moore in a singing role. Rehearsals went ahead and Roger left the cast and there was not much explanation of why. The gag we came up with was that when Lloyd Webber discovered that Roger Moore couldn't sing, he wanted to marry him. It required you to know that he had married Sarah Brightman, the star of his show *Cats*, and that there had been snide remarks in the press about her singing ability. The idea of Roger Moore, the man who played James Bond, being pursued by Lloyd Webber was so absurd, it was funny. Anyway we did this joke once at a party—and within three weeks Lionel Blair had told me that joke and so had Christopher Biggins. And Christopher had told the joke to Simon Fanshawe, who told it to us on air." Kit and the Widow are not convinced that this was just coincidence. They think a little "recycling" had been going on.

Arnold Brown is a stand-up comedian, a profession of eggshell egos and fierce competition for the most original topical gag. Paranoia about having material stolen is a professional inevitability.

Says Brown, "Comedy is about searching for new ideas—it's almost like a scientific process. Suddenly you find that little Rubik's Cube combination—a DNA of comedy which no one else has got to." So when he hears one of his jokes being told by another comedian, does it make him want to sue or does he put it down to coincidence? "Neither," says Brown. "It makes me want to kill them."

Arnold Brown believes he was the first to come up with the comic idea that cell phones were a godsend to the mentally ill, as they could wander around in public talking to themselves and no one would take any notice. But before he could use it in his act, he heard that another comedian was telling an identical joke. Mysteriously, that comedian was never heard of again.

Recently the joke appeared again, this time in Martin Amis's 2003 novel *Yellow Dog*.

What does he put this down to? Coincidence? Great minds thinking alike? Or do jokes get stolen?

"I'm open . . . ," Brown says, "to litigation."

Before he rushes to court, Arnold might take note of the fact that the joke has also been attributed to Jerry Seinfeld as early as 1993. Several other American stand-up comedians also claim ownership. Clearly great comic minds think alike.

Brown thinks the "coincidence" of jokes turning up in other people's acts will continue until someone invents a device that can be placed in a gag, which will explode if it is told by another comedian.

Some of the coincidences that come before the courts are no laughing matter.

Sally Clark was accused of the murder of her two baby sons. The policeman's daughter, who had always protested her innocence, was jailed for life in November 1999. She was convicted of smothering eleven-week-old Christopher in December 1996 and shaking eight-week-old Harry to death in January 1998 at the home she shared with her husband Stephen.

The crux of the case revolved around whether it was conceivable the "crib deaths" of Mrs. Clark's two children were coincidences.

The prosecution's expert witness, an eminent pediatrician told the court that the likelihood of two siblings dying of SIDS or "sudden infant death syndrome" was 1 in 73 million. This was damning evidence against Mrs. Clark and must have had a powerful influence on the jury.

However, on January 30, 2003, after serving three years in jail, Sally Clark won her appeal and freedom. The convictions were ruled unsafe, as medical evidence that might have cleared her was not heard during her trial. The court also criticized the use in the trial of the statistic putting the chance of two babies in the same family suffering SIDS at 1 in 73 million. It said it had been "grossly misleading," as the jury started from the incorrect assumption that double crib deaths in a single family were extremely rare. The assumption had been that the cribs deaths were independent events, and so the seventy-three million figure would have been reached by squaring the probability of a single crib death. But multiple crib deaths in one family are not statistically independent. Experts told the appeal court that the risk of a second crib death could, in fact, have been as low as 1 in 100. It was suggested that the prosecution's expert witness had made a fundamental mathematical error.

The court had decided that whatever the odds, 73 million to 1, or 100 to 1, the deaths of Sally Clark's two children had been natural. The fact that she had lost two children was just a tragic coincidence.

Just months after Sally Clark's acquittal the trial began of another woman accused of the multiple murder of her infant children—a trial at which the pediatrician was again arguing that the deaths could not be the result of simple coincidence.

Thirty-five-year-old pharmacist Trupti Patel denied killing her sons Amar and Jamie and daughter Mia between 1997 and 2001—none of them survived beyond three months. She denied she had smothered her babies or restricted their breathing by squeezing their chests.

In Britain, approximately six hundred children each year die suddenly and unexpectedly at some time between their first week of life and their first birthday. In half of these cases, a clear medical reason for the death is found at postmortem—the remaining, unexplained cases are recorded simply as sudden infant death syndrome.

At the trial of Trupti Patel, the pediatrician stated that "two crib deaths is suspicious, three is murder—unless proved otherwise."

The members of the jury, this time, were not convinced. On June 11, 2003, at the end of the six-and-a-half-week trial, they found her not guilty on all three counts of murder. The jury had decided that the deaths of the three babies, as in the case of Sally Clark, had been a tragic coincidence. Whatever the odds against something happening, the fact that odds can be calculated means that it can, and, given enough time, will happen.

Odds of 73 million to 1, although inaccurately applied in the case of Sally Clark, will, eventually, come up. Even if those had, in fact, been the odds against the deaths of her three children being coincidence, it would not have pointed unerringly to her guilt. A 73 million to 1 chance occurrence isn't an unimaginable likelihood. If one in every seventy-three million people were green, then there'd be eighty-four green people in the world. Shouldn't be too hard to spot.

Judges and juries are regularly asked to weigh up odds of 1 in 3 million—in cases where DNA samples are presented as crucial evidence.

And, of course, they get it wrong.

In 1990, Andrew Deen was sentenced to sixteen years in jail for raping three women. The main evidence linking Deen to the attacks was the close match between DNA samples found at the scene of the crimes and those from Deen. At the trial, the forensic scientist presenting the DNA evidence said that the match was so good that the probability of the samples having come from someone other than Deen was 1 in 3 million. In his summing up, the judge told the jury that so large a figure, if correct, "approximates pretty well to certainty." There could be no coincidence.

But on appeal the court quashed the conviction, declaring the verdict unsafe. It decided that both the forensic scientist and the judge had fallen into a trap known as the "prosecutor's fallacy." They had assumed that the DNA evidence meant that there was only a 3-million-to-1 chance that Deen was not guilty. But they were mistaken.

For the true picture, the appeal court judges turned to a mathematical theorem constructed by a nineteenth-century cleric. Bayes' Theorem addresses the laws of "inverse probability." It provides a formula for working out the impact of new evidence (like DNA samples) on the existing odds of guilt or innocence prior to the introduction of the new evidence.

If "prior probability" of guilt is small—if there is little other evidence to corroborate the DNA evidence—then even the impressive probabilities of genetic fingerprinting can be dramatically diminished.

Researchers at the Institute of Environmental Health and Forensic Sciences in Auckland, New Zealand, use crime statistics and "Bayesian reasoning" to estimate typical prior probabilities. They found that even a DNA match with odds of millions to one can be cut down to final odds against innocence of just 3 to 1—leaving plenty of room for "reasonable doubt."

So if you are currently stuck in an intractable legal dispute over the probability of something or other having happened, or not happened, as the result of pure coincidence—help is at hand. Try applying Bayes's handy mathematical formula.

$$P(A_n \backslash B) = \frac{P(A_n)P(B \backslash A_n)}{\sum_i P(A_i) \, P(B \backslash A_i)}$$

Good luck.

LUCK OR
COINCIDENCE?

It's the day of the Kentucky Derby and you grudgingly hand over your hard-earned $20 for the office pool. Your horse, which has begun with a moderate chance of success, mysteriously chooses to stop half way around the course to admire the stamina and athleticism of its four-legged friends. Your colleague George Robertson wins the jackpot. His horse, a rank outsider, confounds the bookies' pessimistic expectations. This is the seventh time George has won the sweepstake in ten years.

Do you say, "Well done George, it's good to see that the rules of probability are still functioning and that your chances of winning this year were not materially diminished by the fact that you have won so many times before."

Like hell you do. You say, "You lucky s.o.b., George. The drinks are on you."

It's hard not to conclude that someone or something is smiling down on the likes of George Robertson, singling them out for good fortune—leaving the rest of us to muddle along the best we can.

Everything George touches turns to gold. If a nice business trip to Bermuda is in the offing, George gets to wear the shorts. If a promotion is up for grabs, George grabs it. As we know, he wins the Derby pool every year. He won a tidy sum on the football pools a few years back and has even picked up a couple of thousand on the lottery. He's got a beautiful wife, two well-adjusted, respectful kids, a terrific house (bought outright with an unexpected inheritance) and a luxury car. Yes, George is, indeed, among the luckiest of lucky s.o.b's.

But not the luckiest.

Donald Smith of Amherst, Wisconsin, won the state's Super Cash game three times. On May 25, 1993, June 17, 1994, and July 30, 1995. He won $250,000 each time.

Joseph P. Cowley won $3 million in the Ohio lottery in 1987 and retired to Boca Raton, Florida. Six years later he played the Florida Lotto on Christmas Day—and won $20 million.

In 1985 Evelyn Marie Adams won $4 million on the New Jersey Lottery. Four months later she entered again and won another $1.5 million.

Why isn't luck more evenly distributed? What special qualities or mysterious powers are possessed by those few, those lucky few, upon whom Dame Fortune invariably smiles? Is the inordinate amount of good fortune experienced by these lottery winners the result of simple coincidence, or were they born lucky?

What on earth made gambling-mad Mick Gibbs think he could ever pull off the outrageous wager that finally netted him $912,000 in what has been described as the greatest betting coup of all time?

Fifty-nine-year-old Mick of the UK, placed a 50 cent stake on a fifteen-part accumulator bet on who would win a long series of games across four different competitive sports.

*The first fourteen parts of his accumulator all came good, defying stag-
gering cumulative odds. The final part of the wager—that the German team
Bayern Munich would win a European soccer championship at odds of 12 to
1 looked like a long shot. When the game was played on May 23, 2001,
Bayern's opponents, Valencia, looked set to upset the apple cart when they took
a 1–0 lead. Victory for the Spanish side would have meant Mick earned
nothing. Bayern managed to tie before the end of the match and the teams had
to play extra time.*

*Mick was on the edge of a nervous breakdown, pacing up and down in
his garden, unable to watch the match. The game—and Mick's bet—was finally won in the last minute. Bayern won the cup, and Mick won close to
$1 million.*

Mick doesn't put his success down to luck or coincidence. He
believes he won the money because of science. He says he spends
hours poring over the latest sports news and working out his complicated bets.

But if all it takes to win a small fortune is a bit of hard work and
the appliance of a little science, why aren't the rest of the world's ha-
bitual gamblers driving around in flashy sports cars, instead of cy-
cling to collect their unemployment checks?

Can science explain the apparent extraordinary luck of Englishman Charles Wells, the man who broke the bank at Monte Carlo?

Wells's legendary success did not appear to have involved the use
of any system. He walked into the casino in July 1891 and began putting even money bets on red and black, winning nearly every time.
When his winnings passed the one hundred thousand francs mark, the
"bank" was declared broken, the table was closed and a black "mourn-
ing" cloth placed over it. Wells returned the next day to repeat his extraordinary achievement, to the amazement of the casino attendants.

The third and last time Wells appeared at the casino, he placed
his opening bet on number five at odds of 35 to 1. He won. He left
his original bet and added his winnings to it. Five came up again.
This happened five times in succession. The bank had been broken
yet again.

Extraordinary things do happen in gambling casinos. Evens once came up twenty-eight times in succession at a Monte Carlo casino—against odds of 268 million to 1. But was Wells's good fortune simply the laws of probability kicking in? Was he the world's luckiest man? Or was something else going on?

Wells did not get to enjoy his winnings for long. His luck, or whatever it was, dried up. He got involved in a number of shady deals, was arrested by the French police and charged with fraud. Extradited to Britain, he stood trial and was discovered to have had twenty aliases—his real name was never discovered. He was sentenced to eight years in prison. After his release he went to live in Paris where "the man who broke the bank at Monte Carlo" died in poverty in 1926—a broken man.

The secret of Wells's amazing achievements at the roulette table was never discovered. It seems unlikely that his gambling feat was the result of pure luck or, indeed, guided by some supernatural force. Although inspiration for gambling success can come from some pretty strange sources.

> On September 15, 1948, a New York–bound commuter train plunged into Newark Bay killing a number of passengers. Front page newspaper photographs showed the train being winched back out of the water. The number 932 could clearly be seen on the side of the rear coach. Dozens of people took this to be a sign that the number had some sort of significance and chose it in that day's Manhattan numbers game. The number 932 duly came up, winning hundreds of thousands of dollars for the people who had bet on it.

The good luck experienced by fifteen members of the church choir in Beatrice, Nebraska, didn't bring them fame or fortune—it saved their lives.

> Choir practice at the West Side Baptist Church in Beatrice always began at 7:20 on Wednesday evening. At 7:25 P.M. on Wednesday March 1, 1950, an explosion demolished the church. The blast forced a nearby radio station off the air and shattered windows in surrounding homes.
>
> But by an incredible coincidence every one of the choir's fifteen members

*escaped injury. Normally punctual, that evening they were all, and for dif-
ferent reasons, late for practice.*

*The preacher, Walter Klempel, lit the furnace at the West Side Baptist
Church and then went home for dinner. His return to the church with his
family was delayed when his daughter's dress was soiled and his wife had to
iron another for her.*

*Ladona Vandergrift, a high school student, was having trouble with a
geometry problem. She decided to solve it before leaving for choir practice.*

*Royena Estes couldn't get her car to start, so she and her sister called
Ladona Vandergrift and asked her to pick them up. But Ladona was still
working on her geometry problem, so the Estes sisters had to wait.*

*Marilyn Paul, the pianist, had planned to arrive half an hour early, but
fell asleep after dinner . . .*

*And so the list of delays went on. The entire choir, all of whom were
normally punctual for practice, were late that evening.*

*At 7:25, the church blew up. The walls fell outward, the heavy wooden
roof crashed to the ground. Firemen thought the explosion had been caused by
natural gas that had leaked into the church from a broken pipe outside and
been ignited by the fire in the furnace.*

A major tragedy had been averted by the narrowest of multiple
squeaks. The grateful Beatrice church choir members put their amazing good luck down to an act of God. But you don't have to be in a
church choir to narrowly escape death.

*John Woods, a senior partner in a large legal firm, left his office in one
of the Twin Towers of the World Trade Center in New York seconds before the
building was struck by a hijacked aircraft. It wasn't his first close brush with
death. He had been on the thirty-ninth floor of the same building when it was
bombed in 1993, but escaped without injury. In 1988 he was scheduled to
be on the Pan-Am flight that exploded above Lockerbie in Scotland, but canceled at the last minute in order to go to an office party.*

Unlike John Woods, Yugoslavian flight attendant Vesna Vulvic
failed to avoid traveling on a plane destined to explode.

A terrorist bomb was thought to be the cause of the massive explosion
that ripped apart the DC-9 aircraft traveling over the former Czechoslovakia
on January 26, 1972.

Rescue workers who came upon the tangled wreckage on the ground
didn't believe anyone could still be alive. Then they found flight attendant
Vesna Vulvic inside part of the fuselage. She was badly injured, but still
breathing. She was the only person to survive.

Asked afterward how she accounted for her incredible good luck she said,
"I believe we are masters of our lives—we hold all the cards and it is up to
us to use them right."

It's hard to decide whether Vesna Vulvic is a lucky or unlucky
person. She had the misfortune to be on the plane in the first place,
but was extraordinarily lucky to have survived. Just as some people
seem to get more than their fair share of good fortune, others appear
to attract atrocious bad luck.

Frenchman Alain Basseux, a laboratory technician working in England,
lost his temper when a motorist cut him off in a traffic circle. He chased the
offending car for two miles, forced the vehicle to the side of the road, yanked
open the door, and grabbed the driver by the shirt. At this point he noticed that
the man he was assaulting was his boss.

The local magistrates conditionally discharged him for two years, after
his lawyer told the court that such behavior was not unusual in France.

Mr. Basseux kept his job.

So he was lucky in the end. Not so this gentleman:

Businessman Danie de Toit made a speech to an audience in South
Africa warning them that death could strike them at any time. At the end of
the speech he put a peppermint in his mouth, and choked to death on it.

A *New Yorker* cartoon by Mischa Richter pictures God hurling
thunderbolts down from the clouds. "If you're so good," an angel is
saying, "why can't you strike twice in the same place?"

In fact lightning does strike the same place more than once.

> *The Primarda family of Taranto, Italy, has lost three men to lightning strikes, in three generations; two were struck in the same backyard. In 1899 a bolt of lightning killed a man as he stood in his backyard in Taranto, Italy. Thirty years later his son, standing in the same spot, was also struck and killed by a bolt of lightning. On 8 October 1949, Rolla Primarda, the grandson of the first victim and the son of the second, became the third member of the family to step into that garden during a storm.*

Lightning can also strike the same person twice. You might think if you had been struck by lightning once you would have paid your dues to bad fortune and would be invulnerable, yet the odds of being struck again are exactly the same.

A Virginia forest ranger was pursued by lightning with apparently vindictive single-mindedness.

> *During his thirty-six years as a ranger, Roy Cleveland Sullivan was struck no less than seven times. On the first occasion, in 1942, he escaped with the loss of the nail of his big toe. Twenty-seven years later he was struck by a bolt of lightning that singed his eyebrows. The next year, another strike burned his left shoulder. In 1972 lightning set his hair on fire. In 1973 he was blasted out of his car. The sixth strike, in 1976, injured his ankle, and the seventh strike in 1977, while he was fishing, sent him to hospital with chest and stomach burns.*

We expect déjà vu to give us a frisson, not a ten-megawatt charge. What had Sullivan done to deserve such terrible luck? Six years after the seventh strike he committed suicide. The reason, reported at the time, was because he was unlucky in love. A case of *not* being chosen.

Lightning struck Jennifer Roberts only once, as she lay in a tent one night back in October 1991. But once was enough.

> *Caught in a violent electrical storm, twenty-four-year-old Jennifer was struck by a bolt of lightning that scorched the length of her body. She was*

saved from more serious injury because she had just removed her under-wired bra, which had been irritating her. The metal wires would have deflected the electricity to her heart.

The bolt also destroyed the book Jennifer was reading. Its cover bore the image of a head surrounded by lightning flashes.

Good and bad luck often come in clusters. Gamblers talk about being on a roll. Charles Wells's lucky roll, if that's what it was, broke the bank at Monte Carlo. More often the luck runs the other way and the shirt is lost from the gambler's back. We can be blessed with good fortune or cursed with ill-fortune—jinxed. The latter seems to have applied in the case of this royal wedding:

The wedding day of Princess Maria del Pozzo della Cisterna and Amadeo, the Duke d'Aosta, son of the king of Italy, in Turin, on May 30, 1867, was certainly not the happiest day in the lives of a number of those involved. The wardrobe mistress hanged herself, the palace gatekeeper cut his throat, the colonel leading the wedding procession collapsed from sunstroke, the stationmaster was crushed to death under the wheels of the honeymoon train, the king's aide fell from his horse and died and the best man shot himself.

Apart from that it was a lovely day. A similar sort of jinx seems to have afflicted many of the people associated with the comic book character Superman.

The bad luck began with the two men who created the superhero back in 1938. Writer Jerry Siegel and illustrator Joe Shuster signed away their rights to the Man of Steel for a pittance; their various attempts to sue the publishers for a fairer share of the millions made from their creation all failed. Shuster had become a recluse by the time he died.

The actor Kirk Alyn, who played Superman in the 1940s Saturday matinee serial, claimed that it had ruined his career. He struggled to find work and eventually gave up acting. George Reeves, who starred in television's The Adventures of Superman in the 1950s, also struggled professionally when the hit series finished after six years. In 1959, at the age of forty-five,

he was found dead from a single bullet wound to the head. The official verdict
was suicide, but friends were convinced he was murdered.

Christopher Reeve, who played Superman in four films in the 1970s
and 1980s, was thrown from his horse in 1995, broke his neck and ended
up on a respirator and in a wheelchair. Margot Kidder, who costarred as Lois
Lane in all four of Reeve's Superman films, damaged her spinal cord in
1990 in a car accident while filming a TV series and was confined to a
wheelchair for two years. A history of drink and drug abuse and mental ill-
ness eventually led to a nervous breakdown. Richard Pryor, who costarred in
Superman III, was struck down with multiple sclerosis soon after filming
was completed.

Most of us don't experience extremes of either good or bad luck.
Most of us, alas, don't win huge amounts on the lottery and very few
of us fall foul of the "Curse of Superman."

Nevertheless we each tend to think of ourselves as being innately
lucky or unlucky people. We're either the type for whom the bread always falls butter side up . . . or the type whose bread falls, relentlessly,
sticky side down. Lucky people appear to have an uncanny ability to
be in the right place at the right time and enjoy more than their fair
share of life's breaks. Unlucky people are always out when opportunity knocks.

Psychologist Dr. Richard Wiseman has spent the last ten years
finding out why some people lead happy, successful lives while others
face repeated failure and sadness. He also wanted to know whether or
not unlucky people can do anything to improve their luck. He
doesn't believe that good and bad luck is simply a matter of chance.

He says, "For over a hundred years psychologists have studied
how our lives are affected by our intelligence, personality, genes, appearance, and upbringing—but very little work has gone into examining good and bad luck." The results of his study can be found in
his book, *The Luck Factor.*

He decided to search for the elusive "luck factor" by investigating
the beliefs and experiences of people who considered themselves to
be innately lucky or unlucky. The research involved extensive inter-

views with hundreds of people. Many interviews were videotaped and Dr. Wiseman studied not just what his volunteers said but how they said it—their general demeanor.

Lucky people, he noticed, smile more and engage in more eye contact. Lucky people engage in three times as much open body language as unlucky people.

Lucky people persevere with Chinese puzzles, he observed. Unlucky people discard them in seconds, convinced they could never solve them. Lucky people, given a newspaper and told to count the number of photographs, spot the half-page message on page three declaring:

LOOK NO FURTHER,
THERE ARE 42 PHOTOGRAPHS
IN THIS NEWSPAPER.

Unlucky people plough on through to the end, oblivious to the opportunity to curtail their task.

In another experiment Dr. Wiseman took two volunteers who both worked in the business world—Robert (who considered himself a lucky person) and Brenda (who thought herself unlucky)—and invited them, separately, to go to a café and wait for someone connected with the experiment to contact them.

Dr. Wiseman had already placed three stooges at each of three tables in the café. At the fourth table he placed a genuine businessmen who was, potentially, a useful contact for either Robert or Brenda. He wanted to know which of them would manage to take advantage of this real opportunity.

Robert arrived at the shop, ordered a coffee and sat down next to the successful businessman. Within minutes, he'd introduced himself to the stranger, and offered to buy him a coffee. The man accepted and a few moments later the two of them were chatting away.

When it was Brenda's turn, she entered the café, ordered a coffee and sat down next to the businessman. Unlike Robert, she didn't say a word.

The result was much as Dr. Wiseman had anticipated. "Same opportunities—different lives," he points out.

Luck, he concludes, has little to do with chance or coincidence. Lucky people create, notice, and act upon the chance opportunities in their lives. They use their intuition and gut feelings.

"We make our own luck," says Dr. Wiseman. "Your future isn't set in stone. You are not destined always to experience a certain amount of good fortune. You can change. You can create far more lucky breaks and massively increase how often you are in the right place at the right time. When it comes to luck the future is in your hands."

Many of Dr. Wiseman's volunteers believed themselves to be innately lucky or unlucky. "One person, who wanted to be a freelance writer, turned up at a newspaper office just as its regular writer was leaving. She got the job. Her whole life was like that. Another woman had eight accidents in one fifty-mile journey. She put it down to bad luck—and then we saw her park her car and realized that there was more to it than a jinxed car. Another was unlucky in love. Her blind date came off his motorbike and broke his leg. Another prospective boyfriend walked into a glass door and broke his nose. She eventually got engaged to be married, but the church was burned down by arsonists a week before the ceremony."

Dr. Wiseman tested his luck theories by putting many of his "unlucky" volunteers through "luck school," which involved one-on-one counseling sessions, puzzle-solving experiments, questionnaires, and diary keeping. It was all focused on getting people who believed they were unlucky to start thinking and behaving like lucky people. He urged them to change their attitude to bad luck, to trust their intuition, and to spot and take advantage of opportunities when they arose.

He says, "At the beginning we had no idea if it would work. But we have now seen that 80 percent of people feel happier, more satisfied, and most important of all, luckier. We know it works for most people. For some people the improvements are relatively small. For others, especially the very unlucky people, it can have a dramatic effect on their lives. Like Tracey Hart."

Before she went to luck school, Tracey considered herself an exceptionally unlucky person. "Bad luck didn't come in threes for me," she says. "It came in fifteens and twenty-ones." She fell down holes, suffered concussion and a variety of cuts and bruises. A play session with her daughter once laid her out for six weeks. "If I won $20 on the lottery, twenty things would go wrong the next week," she says.

The two lengthy relationships of her life, with the fathers of her two children, both ended in domestic violence. Not surprisingly her health suffered and she became depressed.

And then she met Richard Wiseman and signed up for luck school.

Since then, she reports, she has become a different person. When misfortune strikes now she reminds herself that "it could have been worse." She says she's become a lot more positive in her whole attitude to life. And, indeed, misfortune has become a less regular caller. She can't remember the last time she fell down a hole or had concussion. She's got a new job, a new home, and a new man in her life.

And she's even winning on the lottery and at bingo. "Dr. Wiseman says this is nothing to do with my gaining control of my life. But it's uncanny how often I win now," she says. "A friend of mine was very skeptical about luck school but I persuaded him to follow the principles for a week. During that week we went to bingo and won $1,500 between us. He went home with $1,000 and is now a lot less skeptical."

Dr. Wiseman has recently extended his research to explore whether his luck school principles can be applied to groups of people—in the workplace.

The workplace in question belongs to Technical Asset Management, who repair, upgrade, and recycle computers and other IT hardware. He was invited by Managing Director Kevin Riches to see if he could improve the luck of the company.

Kevin explains, "We'd had an enormous bad debt. It amounted to over three-quarters of a million dollars. To make matters worse we had just gone into a period of expansion and had moved into expensive new premises. The bank found out about the debt and decided to add to our bad luck by withdrawing facilities. We were really struggling. We really needed help if we were to turn things around."

Richard Wiseman came along to their premises and addressed all thirty-eight members of the staff, many of whom were decidedly skeptical that their collective luck could be improved. But nearly all agreed to take part in one-on-one sessions with the psychologist, and to keep diaries that cataloged their progress.

"I know it all sounds very impractical," says Kevin Riches. "Some of my business colleagues thought we were mad and even our chairman said it was all mumbo jumbo. He didn't want it introduced to the rest of the group. Mind you everybody said if it works, let us know."

And did it work?

"It did work. It's been amazing. Productivity went up 20 percent month after month during the period Richard was working with us. And it has continued. We had a record month last month and the business is going from strength to strength."

But could the improvement in the company's fortunes have been just coincidence? "It could have been," says Kevin. "However, working in this business and looking at how we handled new opportunities, I would say our luck training has made a big difference. We have developed a different mindset. We've taken the blinkers off and are now alert to new openings. We've won lots of interesting new business, and it's all been great for morale."

So is Technical Asset Management now a luckier company? "Undoubtedly," says Kevin. "And we will keep it up."

Richard Wiseman is very pleased that TAM has managed to turn around its fortunes, but he's reluctant to take too much credit for the company's good fortune. He says, "It was great that they were open to the crazy idea of a psychologist trying to create a luckier company. But it is just one case study. It would be interesting to see what would happen if other organizations went through the same processes. If we saw the same results again and again we could be more certain that they were the result of luck school."

Kevin Riches turned to psychology to change his company's luck. Those who thought this approach was mumbo jumbo would have been even more appalled if he had chosen to place his faith in lucky charms or superstitious rituals. Yet countless people seem to be prepared to place an unlikely amount of faith in the power of a rabbit's foot or the efficacy of a carefully circumvented ladder.

The famous physicist Niels Bohr was once asked why he had a horseshoe hanging over his office door. "Surely you don't believe that will make any difference to your luck?" a colleague asked. "No," came the reply. "But I hear it works even for those who don't believe."

Psychologist Professor Chris French told us that story. He has been studying superstitions. He is skeptical that certain ritual behavior or the carrying of lucky charms can have anything at all to do with the luck we get in life. He argues that any apparent correlations are pure coincidence.

Says Professor French, "The reason we believe in these things is because it gives us some sense of control of our lives. What is interesting is that the people who tend toward superstitions about luck usually work in situations of uncertainty. By and large, accountants tend not to be superstitious. But actors, sportsmen, soldiers, sailors, students taking examinations, financial investors, gamblers . . . that's a different matter. Gamblers are the most superstitious. However illusory their belief, they are convinced that throwing dice in a particular way will bring them luck. They wouldn't do it otherwise. However, their behavior is completely counterproductive as it draws them into deeper and deeper trouble."

Sports athletes are another highly superstitious breed. Just watch the batter's box antics of Red Sox shortstop Nomar Garciaparra. Chris French accepts that such rituals can have a positive effect in helping concentration and increasing confidence—but nothing more than that. "If the ritual was prevented, it would certainly have a detrimental effect on performance," he concedes. "The problem is that we can't put these things to the test because people won't agree to not carry out their rituals. The Red Sox might have been even more successful if Nomar tightened his gloves in the dugout. We will never know."

American psychologist, Professor Stuart Vyse, provides the following anecdotal examples of superstitions among U.S. sports stars:

> *Buffalo Bills quarterback Jim Kelly forces himself to vomit before every game, a habit he has practiced since high school. Former star Chuck Persons eats two candy bars before every game: two KitKats, two Snickers, or one of each. Former NFL coach George Seifert does not leave his office without patting a book and must be the last person to leave the locker room before a game. And Wayne Gretzky always tucks the right side of his jersey behind his hip pads.*

In 1967, sociologist James Henslin studied the behavior and beliefs of taxi drivers playing the game craps, which is based upon throwing a pair of dice against a backboard. Among the superstitious beliefs identified by Henslin were the following:

- The harder the dice are thrown, the higher the number rolled
- Rituals such as finger-snapping, blowing on or rubbing the dice, etc., can influence the outcome
- The higher the confidence of the thrower, the more likely the desired outcome
- Dropping the dice will adversely affect performance
- Increasing one's bet will positively affect performance.

Professor French concludes that under conditions of uncertainty any belief that gives a sense of control, even if that sense of control is illusory, is likely to be adopted, maintained, and transmitted to others.

Research suggests that women are typically more superstitious than men. Age doesn't appear to be a significant factor, although some superstitions appear to decrease with age, while others increase or stay the same. Arts students typically show much higher levels of belief than students of the natural sciences. Social science students fall somewhere in between.

Belief that superstitious behavior can influence one's luck was described back in 1989 by Leonard Zusne and Warren Jones as "magical thinking." They defined this as "the belief that (a) transfer of energy or information between physical systems may take place solely because of their similarity or contiguity in time and space, or (b) that one's thoughts, words, or actions can achieve specific physical effects in a manner not governed by the principles of ordinary transmission of energy or information."

B. F. Skinner's classic research into "superstition in the pigeon" was conducted at Indiana University in 1948. Skinner described an experiment in which pigeons were placed inside a box and were presented with a food pellet once every fifteen seconds, regardless of their behavior. After a few minutes the birds developed various little idiosyncratic rituals, such as walking round in circles, bobbing their heads up and down and so on. The pigeons appeared to have concluded that their little routines were causing the release of the food even though in reality there was no relationship whatsoever. Skinner's explanation for this phenomenon was that the accidental pairing of the release of food early on in the process with whatever the bird happened to be doing was enough to reinforce that particular type of activity.

But we wouldn't make the same basic mistakes as a pigeon, would we? It's worth bearing Skinner's experiment in mind while thinking about these further observations by Stuart Vyse:

> Bjorn Borg, the five-time Wimbledon champion, comes from a superstitious family. He and his relatives are known for a variety of personal superstitions, several of which center around spitting. As she sat in the competitors' box during the 1979 Wimbledon final, Borg's mother, Margerethe, ate candy for good luck. When Bjorn reached triple match point against Roscoe Tanner, she spat out the piece she had been chewing—perhaps in preparation for a victory cheer. Before she knew it, Tanner had rallied to deuce. Sensing she had made a mistake, Margarethe retrieved the candy from the dirty floor and replaced it in her mouth. Soon her son had won the championship for the fourth time. Earlier that same year, Borg's father, Rune, and his grandfather, Martin Andersson, were fishing and listening to the French Open final on the radio.

Bjorn was playing Victor Pecci of Paraguay. Borg's grandfather spat in the water, and just at that instant Borg won a point. Andersson continued to spit throughout the match, going home with a sore throat. Borg won in four sets.

It seems that our innate love and respect for coincidences leads us to adopt superstitious behavior simply as a result of the accidental pairing of random pieces of behavior and reinforcing events.

Another experiment was carried out by Professor Koichi Ono of Kyoto University in Japan in 1987, using university students as volunteers. They were led into a cubicle where a counter was mounted on the wall behind three levers. The volunteers were told to try to "earn" as many points as possible—though in fact the points registered were predetermined and bore no relation to any activity by the students. Not all of the students engaged in superstitious behavior, but most did. In the case of one female student, the behavior became quite extreme.

About five minutes into the session, a point delivery occurred after she had stopped pulling the lever temporarily and had put her right hand on the lever frame. This behavior was followed by a point delivery, after which she climbed on the table and put her right hand to the counter. Just as she did so, another point was delivered. Thereafter she began to touch many things in turn, such as the signal light, the screen, a nail on the screen, and the wall. About ten minutes later, a point was delivered just as she jumped to the floor, and touching was replaced by jumping. After five jumps, a point was delivered when she jumped and touched the ceiling with her slipper in her hand. Jumping to touch the ceiling continued repeatedly and was followed by points until she stopped about twenty-five minutes into the session, perhaps because of fatigue.

Why have we evolved in such a way that these strange anomalous behavioral patterns persist? Why do we have a cognitive system that is prone to such systematic errors? Or, to put it another way, shouldn't we be smarter than the average pigeon?

Professor French suggests: "The answer is that in evolutionary terms it probably makes more sense to have a cognitive system that works very quickly and usually produces the right answer rather than one that works more slowly but produces a slightly higher proportion of correct conclusions. Our cognition relies upon a range of 'short-cuts,' technically known as heuristics, which generally lead to the correct conclusion but can, under certain circumstances, systematically lead us astray. We have been so successful as a species precisely because we are good at making connections between events and spotting patterns and regularities in nature. The price we have paid is a tendency to sometimes detect connections and patterns that are not really there."

Christina Richards is a seasoned rock and mountain climber, and instructor.

Climbing is clearly a riskier activity than, for example, accountancy (although creative accountancy can have some unpleasant consequences). But is climbing intrinsically dangerous?

"Of course, we are taking calculated risks," says Christina. "Some climbers do it for the adrenaline rush, but most people, if they have any respect for themselves, take proper precautions. They weigh up the risks of death."

How big an element does luck play in mountaineering?

"Luck is a significant factor in climbing. It's tied up in what we call objective dangers; things we can't control, like rock falls, holds snapping, avalanches. These can cause real problems."

So how do climbers deal with these uncontrollable elements? Are they superstitious?

"A lot are. They tend to do things to make themselves feel luckier, and therefore safer. I go through certain routines with my ropes and other equipment, but mainly I carry a ring on a necklace around my neck. It was given to me by a friend, and I do consider it to be lucky.

"Climbers go through all sorts of ritual behavior when starting on a climb. It helps them to focus, but also makes them feel better. No one will ever blaspheme during a climb—even the complete atheists. In such a dangerous environment, you don't want to push your luck."

What would happen if Christina was about to start on a climb and suddenly realized she had left her ring behind at the hotel?

"It would depend how difficult the climb was. If it was a tough one, I would want everything to be in order. If my gear wasn't absolutely right, or I didn't have my ring with me, I just wouldn't do it. Everything has to be in balance."

Does she seriously believe her ring will help her to get safely up or down a mountain?

"If I was embarking on something of lesser consequence—driving to visit a friend, for example—and I hadn't got my ring then it wouldn't be a problem. But when the potential consequence is death—it just isn't worth taking the risk. It's all to do with the scale of the consequences."

What force does the ring exert on Christina's luck?

"I think it is hard to say sitting comfortably talking about it, but there is a feeling that it will make me more lucky, more successful. As ridiculous as that notion sounds, it doesn't sound so ridiculous when you are hanging from a mountain by your fingertips."

Christina's ambition is to climb one of the world's most dangerous peaks—K2. A disturbingly small proportion of climbers who tackle this monster live to tell the tale. Would Christina's ring help give her the courage to take on such a risky climb?

"It would make a small amount of difference. Though training and the abilities of the other climbers would be more significant. With K2 the objective dangers of weather and avalanche are much greater. I'd love to have a crack at it though. I know the chances of surviving are only fifty-fifty—but if you have to die on a mountain—what a way to die."

Is there a danger that climbers might come to rely too much on lucky charms or on their own sense of being a lucky person?

"One guy I know was phenomenally lucky. Got away with stuff where anyone else would have been dead. He would fall and land on his feet or get to the bottom of a climb and his gear would come away and end up in a puddle at his feet. He was an adrenaline junkie. He would overtake on blind bends and get away with it. He considered himself lucky—right up to the time a rock came away in his hand—an objective danger—and he slid down a mountain and broke his back."

Richard Wiseman points out that the risk of believing too much in one's own luck is not limited to adrenaline junkies.

"Research being done by psychologists has revealed a growing body of people who are relying totally on the prospect of good luck on the lottery as a means of progressing in life. Rather than trying to get a job or seek promotion, they are just sitting back and waiting for their lucky numbers to come up. They are convinced it will happen, so see no point in making an effort in any other areas of their life.

"Putting trust in superstitious beliefs in luck is disastrous. Luck simply doesn't work like that. Research shows that unlucky people tend to be more superstitious than lucky people. Lucky people are generally more constructive about the problems in their lives. Unlucky people invest all their optimism in an outside agency. They have a magical viewpoint of luck. The problem is that all these superstitious rituals, touching wood, and lucky charms, don't work and these people just end up getting even unluckier."

Professor Chris French says research proves that people who are psychologically healthy and think of themselves as lucky people are actually less in touch with reality than depressives.

"The truth is that life really is pretty awful," he says. "The depressives have got it right. The people who don't suffer from depression are the ones who have what we call 'unrealistic optimism.' We give people questionnaires to fill in and ask them what are the chances of certain negative things happening to them—of being run over by a bus or contracting a particular illness. Most people assume that the bad things won't happen to them and the good things will. And the truth is that they are being overly optimistic. The depressives tend to be much more accurate. But by living your life as if you were a lucky person, good things will tend to happen to you because you will be willing to take risks. By not living in an overcautious, worried, anxious way, you will get more out of life. This is a nice example of a situation where an irrational belief can be psychologically healthy."

In the final analysis, whether we go to luck school and learn to control our luck, carry lucky rabbit's feet in the hope of warding off bad luck—or simply leave ourselves open to whatever luck is handed down to us from on high (literally in the case of forest ranger Roy Sullivan), Dame Fortune can be very mischievous.

> *In June 1980, Maureen Wilcox bought tickets for both the Massachusetts and Rhode Island lotteries. She had the winning numbers for both but didn't win a penny. Her Massachusetts numbers won the Rhode Island lottery and her Rhode Island numbers won the Massachusetts lottery.*

What on earth could she have done to deserve that?

. 7 .

DOES COINCIDENCE
ADD UP?

Mathematicians are not fanciful people. They are rationalists, using numbers to understand life's mysteries. Where others see coincidences as evidence of magic or divine intervention, they see the laws of probability in action.

So how unlikely would something have to be, how long the odds against it, before a fusty old mathematician was prepared to accept that it was beyond coincidence—that something really rather strange was going on?

What's the most unlikely thing you could imagine happening? Winning millions on the lottery—twice? Being struck repeatedly by lightning? These things happen, as we have already seen. They don't happen very often, of course, and not usually to us, but they do happen. Mathematicians tell us that if a thing can occur it will occur—eventually. Only impossible things don't happen—like discovering icebergs in the Sahara or taxis in the rain.

How does the mathematics of probability—of coincidence—

add up? What would be the odds, for example, of being struck by a meteorite just minutes after discussing the odds against such a thing happening? And if it happened, would a mathematician be prepared to believe that it was just coincidence? The answer to this a little later.

Meanwhile, at the other end of the probability scale, how surprised should we be when we meet someone at a party who happens to have the same birthday?

With odds of 365 to 1 against, it doesn't seem like this should happen too often. When we find someone who shares our birth date, we tend to think something a bit special has happened. Fancy that, of all the dates in the year, we have the same birthday. What a coincidence!

Surprisingly the (rather complicated) mathematical formula dictates that you only need twenty-three people in a room for there to be a better than 50 percent probability that two of them will share the same birthday.

It seems an absurdly low figure—one worth putting to the test. We looked for an average sample of people. Where better to find one than on the street?

In the event, we had to ask twenty-nine people before we found a match—a young girl waiting for a bus was born on 24 July, the same day as the sixth person we spoke to. The girl at the bus stop was not remotely surprised that we had only had to stop so few people. In fact she thought it odd we had had to ask as many as twenty-nine. Her boyfriend, and four of her friends, all shared her birthday!

Eminent mathematician Warren Weaver once explained this at a dinner attended by high-ranking U.S. military men and then started around the table to compare birthdays. To his disappointment, he reached the last officer without turning up a single coincidence. But he was rescued by the twenty-third person in the room. The waitress, who had been listening, announced that she had been born on the same day as one of the generals.

Mathematical truths are often counterintuitive. The reality can surprise and delight us—or, at times, disturb us. We have a natural tendency to think the likelihood of something happening is either much greater or, indeed, much smaller than it really is. Our underestimation of the odds against winning the lottery keeps us buying tickets, and our overestimation of the odds against a road accident keeps us driving our cars.

Let's look at some other improbable things. If you were playing bridge and received a hand containing thirteen cards of the same suit, you would be amazed. And yet that eventuality is no more likely, or unlikely, than any other combination of cards. The likelihood of receiving any *predicted* hand of cards is, of course, another matter. The odds against being dealt all thirteen spades, for example, has been calculated as 635,013,559,600 to 1.

So the average bridge player shouldn't expect this sort of thing to happen too often in a lifetime, unless he happens to live on the planet described by evolutionary biologist Richard Dawkins in his book, *The Blind Watchmaker*:

> *If on some planet there are beings with a lifetime of a million centuries, their spotlight of comprehensible risk will extend that much farther toward the right-hand end of the continuum. They will expect to be dealt a perfect bridge hand from time to time, and will scarcely trouble to write home about it when it happens.*

When Dawkins says "dealt a perfect bridge hand," he means when someone receives a "perfect deal," such as thirteen cards of the same suit. A "perfect hand" in bridge would be one that cannot be beaten, and involves quite different mathematical calculations. In case you are interested, the odds against being dealt a perfect hand in bridge are 169,066,442 to 1.

Anyway, even on a planet where people live for countless millennia, the prospect of *all four* players in a bridge game receiving perfect deals looks a little unlikely. Dawkins calculates the odds against this happening as 2,235,197,406,895,366,368,301,559,999 to 1.

As mind-boggling as those odds may seem, such an extraordinary event has, apparently, happened—at a whist club in England, back in January 1998. As reported in the paper:

> *Eighty-seven-year-old Hilda Golding was the first to pick up her hand.*
> *She was dealt all thirteen clubs in the pack. "I was amazed. I'd never seen any-*
> *thing like it before, and I've been playing for about forty-odd years," she said.*
> *Hazel Ruffles had all the diamonds. Alison Chivers held the hearts. The*
> *spades were with the dummy. Alison Chivers insists that the cards were shuf-*
> *fled properly. "It was an ordinary pack of cards. They were shuffled before*
> *they went on the table, and Hazel shuffled them again before they were dealt."*

The elderly members of the whist club had just beaten astronomical odds. It was actually more likely that each would win the jackpot in the national lottery and the football pools in the same week. Unfortunately it didn't win them a penny.

So just how astonished should we be by such an event? That we know about it at all is a product of "selective reporting." The newspapers printed the story because they had decided that it had been a remarkable thing. The perfect hand of bridge is more likely to make headlines than an imperfect one. We don't get headlines saying "Bridge Players Are Dealt Random Shuffle of Cards."

William Hartston, author of *The Book of Numbers,* believes we get too excited about coincidences. For example, he was less than impressed by the story of two golfers who hit a hole in one with successive shots. The players had the same surname but were not related. Wasn't this rather extraordinary?

Not a bit of it, says Hartston, "First of all, let's dispose of the little matter of the golfers having the same name. The tournament was in Wales and their shared surname was Evans."

But Richard and Mark Evans had both hit a hole in one at the third with successive shots. What are the odds against that?

Hartston estimates that the chance of a hole in one varies from I in 2,780 for a top professional to I in about 43,000 for a club-swinging amateur. In the latter case he calculates that on any given hole, the chance of two players acing the ball with their tee shots, one after the other, would be 1.85 billion to 1.

But isn't that pretty staggering?

Apparently not, Hartston explains, "If 2 million golfers play an average of two rounds of golf a week each, that's more than 200 million rounds of golf a year, amounting to a total of 3.6 billion holes. That 1.85 billion to 1 shot doesn't look so unlikely anymore, does it?" In fact, if Hartston's calculations are correct, we should expect this sort of thing to happen somewhere about once a year.

He argues that stories and statistics like these show two things: first, that we are bad at assessing probabilities and second, that we tend to err in the direction of optimism. "Encouraged by stories of holes in one, royal flushes, and jackpot wins, we swing our golf clubs in blind hope and gamble our spare cash on impossible odds, hoping to catch the eye of Dame Fortune. Yet at the same time we play sports, where the injuries send millions to the hospital every year, we travel by car, which kills many people every day, and we smoke, which causes hundreds of thousands of deaths a year."

In the fifty years following the first conquest of Everest by Sir Edmund Hillary and Sherpa Tenzing in May 1953, 800 people climbed the world's highest mountain. Of those, 180 died in the attempt. William Hartston points out that the ratio of successes to deaths is roughly 5 to 1—the same odds as Russian roulette.

Accidents and misfortune, we like to think, happen to other people. Acts of incredible good fortune, we hope, will happen to us. Certainly our general inability to fully grasp the subtleties of the laws of probability can lead to some very strange attitudes toward risks in life.

A paper produced by the Said Business School points out that we all regularly run the risk of being killed in a road accident. Almost 1 man in 100 (though many fewer women) dies that way. How much, therefore, would we pay for extra safety features that would halve the risk, such as airbags and crumple zones? A thousand dollars, or perhaps as much as $2,000? But how much would you want to be paid before you would agree to cross a minefield in which there was a 1 in 100 chance of you being killed? Almost certainly more than $2,000, suggests the paper.

Anyone seriously attempting to understand the significance of coincidences (and who wants to be clearer about the relative risks in life) might find the following statistics helpful:

- The odds against winning the U.S. Powerball jackpot with one ticket: 80,089,128 to 1
- Being dealt a royal flush at poker: 649,739 to 1
- Hitting a hole in one with any one shot: 42,952 to 1
- All four players drawing perfect hands of bridge: 2,235,197,406,895,366,368,301,559,999 to 1
- Being murdered in the next year: 18,141 to 1
- Being struck by lightning: 600,000 to 1
- Dying in a railway accident: 500,000 to 1
- Dying under the wheels of a bus: 1,000,000 to 1
- Dying in a plane crash: 10,000,000 to 1
- Choking to death on food: 250,000 to 1

And the odds against two Welshmen having the same surname: 15 to 1.

What are the odds against dreams coming true? Accounts of prophetic dreams have been reported through the ages—by the ancient Assyrians and Babylonians and throughout Egyptian, Greek, and Roman civilizations. There are numerous accounts in the Bible. And they still happen.

Sharon Martens of Milwaukee, Wisconsin, was fourteen when she met and became firm friends with a boy named Michael. About a year later she had a disturbing dream—that she and Michael were at a basketball game and he told her he was leaving town the following Tuesday. Later that week, Michael approached her at school and told her his family had made the sudden decision to move to Colorado. When was he going? The following Tuesday, he told her.

Did young Sharon have some sort of psychic premonition? Or was this just coincidence? And if it was just coincidence, what would be the odds against such a thing happening? In an article published in the *Washington Post* in 1995, Chip Denman, a statistics lecturer at the University of Maryland, worked it out.

He made a series of complex mathematical calculations, involving various assumptions about how often we dream and the odds against any individual dream coming true. He eventually came to the conclusion that the average person, simply as the result of chance and without the help of special psychic powers, will have a dream that accurately anticipates future events, once every nineteen years. "No wonder so many of my students tell me that it has happened to them," says Chip.

Mathematician Ian Stewart of the UK has studied the phenomenon of coincidence. He remains skeptical that the explanation for seemingly impossible chance events lies outside the realm of the laws of probability. He thinks people who assume something paranormal is going on are failing to grasp the facts.

Was it possible, then, to come up with a coincidence story that Ian Stewart could not explain in purely mathematical terms? Professor Stewart was prepared to rise to the challenge. Game on.

Martin Plimmer had been on vacation with his wife and children and they were playing a coin-tossing game, guessing heads or tails. His wife guessed heads or tails correctly seventeen consecutive times. Was that just coincidence?

Professor Stewart was dismissive. "Think about it mathematically. We assume heads and tails are equally likely ... one half times one half seventeen times, that's going to be about ... I in 100,000 probability. This is fairly unusual. Something like that happened to me once. It's just a I in 100,000 chance. Occasionally you get lucky."

That was rather disappointing. Martin's wife hadn't, in some mysterious way, influenced the fall of the coin, or some how "read" how it had landed or attained a rare level of cosmic harmony with her children. She'd just been lucky.

For a really time-consuming holiday distraction she should have tried to flip a coin so it landed heads fifty times consecutively. Apparently to achieve this would take a million men tossing coins ten times a minute, forty hours a week—and even then it would happen only once every nine centuries. But it would happen. And the men could then, presumably, go home.

What did Ian Stewart make of the following coincidence?

At the 1997 Spanish Grand Prix, three racing drivers, Michael Schumacher, Jacques Villeneuve, and Heinz-Harold Frentzen, all lapped in exactly 1 minute 21.072 seconds. Was this not, as the astonished commentators suggested at the time, an extraordinary coincidence?

Again, Professor Stewart was not impressed. "The top drivers all lap at roughly the same speed, so it's reasonable to assume that the three fastest times would fall inside the same tenth-of-a-second period. At intervals of a thousandth of a second, there are one hundred possible lap times for each to choose from. Assume for simplicity that each time in that range is equally likely. Then there is a 1 in 100 chance that the second driver laps in the same time as the first, and a 1 in 100 chance that the third laps in the same time as the other two—which leads to an estimate of 1 in 10,000 as the probability of the coincidence. Low enough to be striking, but not so low that we ought to feel truly amazed. It's roughly as likely as a hole in one in golf."

A man riding a moped in Bermuda was killed in a collision with a taxi, exactly a year after his brother had been killed—on the same street, by the same taxi driver, carrying the same passenger, and on the very same moped.

"This is another one where the chances are low, but the circumstances conspire to make it happen," says Stewart. "The brother was using the same moped, so he obviously wasn't superstitious. It was probably a dangerous street. The taxi driver obviously was not a good driver. This experiment is carried out millions of times every year. You don't hear stories about someone being killed by a different taxi driver. This kind of event is very unlikely, but every so often it will happen."

It was time for Martin Plimmer to unleash his "killer" story.

Martin had taken his six-year-old son to the doctor for a small opera-tion. When the nurse administered an injection, Martin fainted—hitting his head as he fell. The hospital insisted he have an X ray. He arrived at the de-partment and was told to wait. On the table in front of him was a four-year-old magazine—open at an article he had written—on the subject of headaches.

"That's a good one." said Professor Stewart. "It's surprising and unusual. Things like that don't happen to you very often, which is why we find coincidences striking. Given all the factors involved, the odds against it happening must be in the region of a 1,000,000 to 1. But how many things happen to you in a day? A thousand things? At least. Over three years . . . a thousand days of a thousand things a day, a million things happen to you. In among those there will be one whose chances are one in a million. So about once every three years something like that ought to happen to you. If it's hap-pening to you more often than that, then it is getting interesting mathematically."

Professor Stewart says the reason we tend to be so amazed when these coincidences occur is not simply because they occur—but be-cause they happen to us. "Of all the people in the world it could have happened to, it happened to you. The Universe picked you. And there's no explanation for that."

He adds that our intuition is worse than useless when we think about coincidences. "We're amazed when we bump into friends in unusual places, because we expect random events to be evenly distributed—so statistical clumps surprise us. We think that a 'typi-cal' draw in a lottery is something like 5, 14, 27, 36, 39, 45—but that 1, 2, 3, 19, 20, 21 is far less likely. In fact these two sets of numbers have the same probability—1 in 13,983,815. Sequences of six random numbers are more likely to be clumpy than not."

What then did Professor Stewart make of one of the most famous of all coincidence stories—that which connects the lives and deaths of Presidents Abraham Lincoln and John F. Kennedy?

> *Abraham Lincoln was elected to Congress in 1846. Kennedy was elected in 1946. Lincoln was elected president in 1860, Kennedy in 1960. Their surnames each contain seven letters. Both were concerned with civil rights. Both were shot on a Friday. Both were shot in the head. Both were assassinated by men with three names comprising fifteen letters. John Wilkes Booth who assassinated Lincoln was born in 1839. Lee Harvey Oswald who assassinated Kennedy was born in 1939. It goes on and on. . . .*

"If you just take the list of things, it sounds like a very unlikely chain of events. But it's numerology. People are looking for the things that are the same and ignoring all the things that are different. You focus on the fact that some names have the same number of letters but other names don't. How many letters on average would three names have? Well fifteen is probably close to average. The fact that they were born one hundred years apart means their careers are likely to track each other roughly one hundred years apart. Both were shot on a Friday—well there is a 1 in 7 chance.

"If you play these games and look for similarities and are prepared to be imaginative about what you look for and only count the things that are similar, I suspect you can take any two people on the planet and find an amazing amount of things in common.

"The fact is that they are both human beings, which means they have a lot in common to start with. You just have to find out what it is."

And there is evidence to back him up on that. Awhile back *The Skeptical Inquirer* held a competition to find "amazing coincidences" between other world leaders. The winning entry unearthed sixteen uncanny similarities between Kennedy and President Alvaro Obregon of Mexico.

Arthur Koestler suggested that a possible explanation for coincidences is that like things in the universe may be attracted to each other. Did Ian Stewart have any sympathy with that view?

"There is a sense in which that is true," said Professor Stewart. "But for obvious reasons. People who travel by aircraft a lot will be attracted to one another in airports. It's not a surprise if lots of coincidences happen to me in airports, because I spend a lot of time in them.

"As regards a more mystic kind of attraction of like things, I'm not convinced. Some people suggest that there is a secret hidden order to the universe and it is our job as scientists to work it out. But that kind of unity in the universe is on such a deep level—of fundamental particles all obeying the same rules—that it does not translate into anything meaningful on the level of people in terms of an obvious association of like with like. . . ."

And then suddenly, and unexpectedly, a chink appeared in the mathematician's armored skepticism about the existence of some sort of synchronistic force that creates coincidences.

". . . but on the other hand I wouldn't say it was nonsense. I mean the universe is a very strange place and it does function in ways we don't understand very well."

Did this mean that it was, after all, possible to think of a coincidence scenario that Professor Stewart could not dismiss as the result of pure chance—that was "beyond coincidence"?

"The place where I lose confidence in my explanations," he said, "is when I get to the point when I'm not explaining—but explaining away."

What, for example, if we were talking about meteorites and one landed on a nearby building? Could he explain that away?

"I don't think I could. It would be very difficult. At the very least it would be a remarkable thing—an astonishing thing."

And what about the chances of actually being struck by a meteorite?

The chances, he reckoned, were astronomically slim. The only known instance was in 1954 when a nine-pound meteorite crashed through the roof of Ann Hodges's home in Sylacauga, Alabama, and struck her in the hip as she slept on a couch. She escaped with a large bruise.

So did Ian Stewart think it was safe to leave the building?

"Well, the problem is you just don't know. You could go somewhere else and that turns out to be the place that gets hit."

You'll never guess what happened when we walked out of the building . . .

Nothing.

Two

COINCIDENCE ON

THE

RAMPAGE

· I ·

IT'S A
SMALL WORLD

When coincidence taps us on the shoulder in the form of an old friend in a strange place, we marvel at what a small world we live in.

Everyone agrees that the invention of the airplane has made the world an even smaller place. Not so small you could put it in your pocket, perhaps, but small enough to travel halfway around the world in the time it takes to watch half a dozen bad movies.

In fact the world has remained resolutely the same size (7,925 miles in diameter at the equator the last anyone checked), give or take a few quarters of an inch for natural shrinkage. And that's pretty big really, though not as big as Jupiter, of course, which is one thousand times greater in volume. If coincidences occur in direct proportion to the smallness of the planet, then presumably they occur one thousand times less frequently on Jupiter. Someone should look into that.

Anyway, here on Earth, jets certainly enable us to get about a lot more and therefore increase our potential for experiencing coincidences. Our forebears did not have the glorious blessing of package tours to Disneyworld. Their coincidence potential was restricted to the environs of their village or, as this old joke has it, the local bar:

> A man stumbles up to the only other patron in a bar and asks if he can buy him a drink. "Why, of course," comes the reply.
>
> The first man then asks; "Where are you from?"
>
> "I'm from Ireland," replies the second man.
>
> The first man responds, "You don't say, I'm from Ireland, too! Let's have another round and drink to Ireland."
>
> "Of course," replies the second man.
>
> "I'm curious," the first man then asks, "Where in Ireland are you from?"
>
> "Dublin," comes the reply.
>
> "I can't believe it," says the first man. "I'm from Dublin, too! Let's have another, and drink to Dublin."
>
> "Of course," replies the second man.
>
> After a while the first man asks, "What school did you go to?"
>
> "Saint Mary's," replies the second man, "I graduated in '62."
>
> "This is unbelievable!" the first man says. "I went to Saint Mary's and I graduated in '62, too!"
>
> Another customer enters the bar. "What's been going on?" he asks the bartender.
>
> "Nothing much," replies the bartender. "The O'Reilly twins are drunk again."

Neither alcohol nor the miracle of modern aviation can account for many of the extraordinary stories that fall into the "small world effect" category of coincidences. Let's look in a little more detail at the story of the two Laura Buxtons.

In June 2001, ten-year-old Laura Buxton was at a party where she wrote her name and address on a luggage label, attached it to a helium balloon and released it into a clear blue sky.

The balloon floated 140 miles until finally coming to rest in the garden of another Laura Buxton, aged ten.

The second Laura immediately got in touch with the first Laura and the girls have since become friends. They've discovered that not only do they share the same name and are the same age, they are both fair haired, and each owns a black Labrador, a guinea pig, and a rabbit.

Now you may not get that sort of thing happening on Jupiter. But Earth is clearly a very small place—as these following stories confirm.

DÉJÀ-VICKY

R. T. Kallidusjian's ears pricked up when a stranger at a dinner party mentioned that his first wife was called Vicky Bigden. Vicky Bigden was his first girlfriend. Subsequent conversation revealed that the first man had married Vicky at the same time (2 p.m.) on the same day (Saturday July 11, 1964) as Kallidusjian had married his first wife. Both couples attended the Antibes Jazz Festival later that summer.

THE POIGNANT POSTCARD

When James Wilson's father died in South Africa in the 1960s James was on holiday in Spain. He cut short the holiday and made arrangements to fly to South Africa, via the Canary Islands, where he joined up with his brother-in-law in order to continue the journey together. While in the Canaries airport they bought a postcard to send to James's sister in Holland. The picture showed people walking on a beach. One of the people was James's father.

HEY MISTER,
THAT'S ME UP ON THE JUKEBOX

UK musicians The Gibsons recorded three singles on the Major Minor record label in their sixties heyday, and for a while they made encouraging headway on the cabaret circuit. But none of the singles made it into the charts and when the sixties petered out so did they. Years later ex-Gibson Bernie Shaw followed a classified ad to a farm in the middle of rural Lancashire, hoping to buy an old jukebox. The farmer led him across a rutted field to an old barn and there, covered in straw, stood the music machine. It could hold forty seven-inch records, but only two remained, covered in dust. One didn't have a decipherable label; the other was "Only When You're Lonely," by the Gibsons, featuring Bernie Shaw on vocals.

ONE GOOD TURN
DESERVES ANOTHER

Allan Cheek's first promotion in business brought about a rude awakening for him. His employer summoned Cheek to congratulate him on the way he had revived various business operations he ran. The time had come, he said, to involve him in a more trusted and responsible way in his serious business plans.

It wasn't long before Cheek realized just how serious these were. The first plan was to swindle a great deal of money out of a prospective investor. Cheek refused to get involved, telling his boss that if he went ahead with it he would resign and go straight to the intended victim and warn him.

The boss was furious. "You can't afford to be squeamish," he said. He was right: Cheek was hard up and needed the job.

"But I really had no choice," he said. "I resigned." Immediately he drove 180 miles to where the unsuspecting victim lived. The man was a little discomfited by the surprise visit; he didn't like to think he might have been taken in.

"Well, it's up to you," said Cheek. "I've done all I can." And he drove home again.

Two years later he was working for a new company that had got itself into serious trouble. In just a few months the original invest-

ment had been frittered away and it had run up a huge bank overdraft. The chairman decided it was time to pull the plug and fire the
CEO and close down the operation.

Cheek thought something could be salvaged, though, and worked
all night on a report setting out how the company might prosper if it
were allowed to continue. For several hours the next day the chairman
argued with him over the report before finally agreeing to leave
Cheek in charge. "Be it on your own head" were his last words.

Cheek had to work his miracle on a shoestring, so the first thing
he did was to move the company out of the expensive suite of offices
it was renting. But now he needed an office. In the classified ads of the
local evening paper he saw an advertisement for three measly rooms
over a garage. He made an appointment to view them. They were
barely adequate, but they were cheap. Even so he couldn't afford them.
He followed the landlord down the stairs and out into the street.

"They'll do," he said, "but there is one small snag. I can't pay you
any rent—yet." He explained his predicament, hoping wildly that the
landlord might share his faith in his near-bankrupt company.

The landlord was silent for a while, then surprisingly said,
"What did you say your name was?"

"Allan Cheek."

"Did you two years ago warn a man he was about to be swindled?"

"Yes."

"That was my brother. He would have lost his life savings. Move
in when you like and pay me when you can."

Within four years the company had not only repaid its generous
landlord and cleared its overdraft, but it could also afford to move
into a brand-new office park, with a warehouse attached.

THE LATE GUEST

Patti Razey was invited to her friend Janet's wedding. She couldn't go
because she'd already agreed to go on vacation to Tunisia with Liz.

After two days Liz was informed of a death in her family and had to return home. Disappointed at being without friends, Patti decided to make the most of it and go anyway. On the plane was Janet with her new husband. Patti said, "Well I couldn't get to the wedding, but I did make it to the honeymoon."

LOST PROPERTY

In 1953, the *Chicago Sun-Times* columnist Irving Kupcinet checked into the Savoy Hotel, London, to cover the coronation of Elizabeth II. He opened a drawer in his room and found some personal belongings of an old friend, the basketball player Harry Hannin of the Harlem Globetrotters. Two days later Hannin sent a letter to Kupcinet from the Hotel Meurice in Paris. Hannin wrote, "You'll never believe this but I've just opened a drawer here and found a tie with your name on it." Kupcinet had stayed in that particular hotel room a few months earlier.

THE RETURNING MANUSCRIPT

An aspiring writer submitted his manuscript to a publisher and waited anxiously for a response. Some time later he found the manuscript lying in his back garden. Angrily he rang the publisher to ask what on earth was going on.

The publisher explained that she had in fact been very impressed with the work. It had been stolen the previous night from her car along with a number of other things when she was at a restaurant.

She could only conclude that the thieves had not thought as highly of the manuscript as she had, and had tossed it over the nearest garden fence.

ONE GOOD TOURNIQUET

The destinies of Allen Falby, an ordinary El Paso County highway patrolman, and Alfred Smith, an ordinary businessman, came together not once, but twice, to the two men's consecutive advantage.

The first time they met, on a hot June night, Falby was lying in the road in a pool of blood, one of his legs rapidly hemorrhaging blood from a ruptured artery. He had flown off his motorcycle while trying to overtake a speeding truck. The truck had braked without warning and Falby had slammed into the tailgate.

Smith was driving home when he came across the accident. He had no medical training but he could see immediately that Falby was bleeding to death and common sense told him what to do. With his tie he fashioned a tourniquet. An ambulance crew that arrived a few minutes later said it might have been the crucial element that saved Falby's life.

Falby spent months in hospital, but eventually returned to work. Five years later, at Christmas, Falby was on highway night patrol when he received a call to investigate an accident on U.S. 80, in which a car had run into a tree. Falby, who was first on the scene, found Smith slumped unconscious in the car, bleeding profusely from a severed artery in his leg. Falby, who was trained in first aid, quickly applied a tourniquet above the ruptured artery. As he put it later: "One good tourniquet deserves another."

STRANGERS ON THE SHORE

A couple basking on a beach caught John Peskett's eye as he flicked through his wife's old photographs of a childhood holiday. He looked closer—and realized to his astonishment that the couple sunbathing were his parents.

John and his wife-to-be Shirley, then both aged ten, had been just feet apart on the same sands at the same time without realizing they were destined for each other.

They grew up hundreds of miles apart, but in 1963 both their families spent their summer vacations on the same bit of beach, paddled in the same bit of sea, and went home without making contact.

In 1974 John and Shirley met at a teacher training college and began dating. The romance led to marriage.

When Shirley brought out her snapshot of the vacation, John first spotted a duffle bag and a football he had been playing with that day.

He said, "That woman there looks just like my mother."

"Then I did a double take and realized she was my mother. We had the picture blown up and took it to show my parents.

"They were suitably shocked, but they were also great believers in fate and thought we must have been meant for each other."

John said, "I believe in fate—I think it puts you in the right place at the right time."

PROPHETIC SIGN

Eileen Bithell was so astounded by this coincidence that she felt compelled to write to the *Times* about it.

"For over twenty years there hung in the window of my parents' grocery shop a framed sign indicating which day of the week the shop closed. Two weeks before my brother's wedding, the sign was taken down to be altered and was removed from its frame. On the back of the sign was a large photograph showing a small girl in her father's arms. The man was then the mayor of the town and was officiating at the opening of a new hospital. The small girl was my brother's bride-to-be and the man his future father-in-law. No one knows how this particular photo came to be used as the backing for our shop sign as none of the people in the photo were then known to my family, yet now, twenty years later, the two families were to be joined in marriage."

ELECTRIFYING COINCIDENCE

The distinguished poet Craig Raine recalls being asked by composer Nigel Osborne to write the libretto for the opera, *The Electrification of the Soviet Union*.

"Nigel rang me up one evening and said, 'Are you interested in writing the opera?' I thought about it for fifteen seconds and said that was a very good idea. I then asked him if he had an idea and he said he had in mind a story by Boris Pasternak called *The Last Summer*.

"I said, 'You are not going to believe this but I took it down off my shelf yesterday thinking I would read it again. I can actually see it from here.'

"He said, 'The only trouble is, we might not get permission to do it from the Pasternak estate.'

"I said, 'I don't think we'll have too much trouble there. My wife is Boris Pasternak's niece.'"

A CHANCE ARREST

A series of coincidences enabled police to arrest a man believed to have been responsible for about two dozen armed robberies of San Fernando Valley businesses. In most cases the robber entered stores brandishing a gun. He forced staff to lie on the floor, stole cash, and made his escape.

Police made little progress tracking down the culprit until, by chance, a woman visiting the Los Angeles Police Department's West Valley Division on other business happened to see a composite drawing of the robber. She identified him as Douglas McMann.

Desk officer Ken Knox ran a computer check and found that police had briefly held McMann in custody for several days for an unrelated crime. The next day Knox happened to be walking along Roscoe Boulevard when he spotted McMann and arrested him. He was charged with three counts of robbery and one count of assault with a deadly weapon.

.2.

IT'S A TOO
SMALL WORLD

Just as the "small world effect" can bring us wonderful surprises, welcome reunions, rescue from imminent death, and other benefits, it can also bring us pain, humiliation, and even the unwanted attention of the police. That's when our small world becomes too small a world.

Many of the people caught up in the following stories would surely wish to have been born on a much larger planet.

TWO SISTERS

Two sisters driving separate vehicles on a rural highway collided head-on, and were both killed. They were traveling to see each other. State troopers said that Sheila Wentworth, forty-five, and Doris Jean Hall, fifty-one, were driving Jeeps in opposite directions on Alabama

25 when one of the vehicles crossed the median strip and collided with the other vehicle.

WRONG NUMBER

Amanda recognized the voice on her answering machine, even though the caller had dialed the wrong number. What's more the male voice recognized her voice. "I misdialled but I know your voice, even though we haven't spoken in fifteen years," he said. "I've listened to the message a dozen times now and I know it's you."

Amanda had thought that John was out of her life. Fifteen years before they'd had a passionate relationship, but he had broken it off for another woman, and now had three children. Amanda had also married since and had four children.

John left another message asking her to call him. She did, and then she went to see him. Now Amanda and John are once again trapped in a love affair, desperately trying to come to terms with the potential consequences for their families.

COINCIDENCE, OR NOVELTY VENDETTA?

A stray golf ball hit a man on a course in the UK. Ten days later his wife was hit by a ball at the same spot, played by the same golfer.

CHECK MATE

Vincent Leon Johnson and Frazier Black picked the wrong bank teller when they tried to cash a stolen check.

A court in Austin, Texas, heard how the two burglars broke into the home of David Conner, helping themselves to two color television sets and checkbooks belonging to Conner and his girlfriend Nancy Hart—who were both at work at the time.

A few hours later, Johnson and Black showed up at the bank where Nancy worked as a teller. They presented her with a forged personal check for $200 belonging to Hart and made out to Conner. Bank security staff held the men until the police arrived.

KARPIN'S CAR

French intelligence agents arrested a German spy, Peter Karpin, in France, shortly after the outbreak of the First World War. They kept the arrest secret and sent fake reports from Karpin to his superiors, at the same time intercepting money sent on his behalf back to France. The funds were used to buy a car. Karpin escaped in 1917. Two years later, after the war was over, the car ran down and killed a man in the French-occupied German Ruhr. The victim was Peter Karpin.

TILL DEATH

Two cars collided at high speed in Paris in 1996 killing both drivers—who turned out to be man and wife. The couple had been separated for some months and neither knew that the other would be out driving that night. Police considered the possibility of it being a bizarre murder-suicide but concluded it was pure coincidence.

THE WRONG BAG

Police in Italy caught a thief after he sped past a woman on his motorcycle and snatched her purse. The woman was his mother, who recognized him and reported him.

COLLARED

A thief thought he should look his best when he appeared in court on burglary charges. So he wore his smartest jacket.

But it backfired when the prosecutor assigned to the case, Marc Florens, recognized the jacket as his own, stolen from his home earlier in the year, along with a camera and some cash. Florens retrieved his jacket, but as an "interested party" he had to be replaced by another lawyer before the case could proceed.

DON'T PLAY THAT SONG TO ME

Peter Robertson's neighbor Sarah turned up at his apartment in a distressed state because of intractable problems with her boyfriend. Tactfully, Peter put some gentle music on the CD player to soothe her, but the first note of Van Morrison's "Someone Like You" made Sarah wail. It was her and her boyfriend's special song.

OUT-SWISSED

There was only one soldier in the Confederate Army during the American Civil War who understood Swiss, and it was just the luck of the Swiss prisoners captured fighting for the Union that it was his turn on guard duty the night they plotted their escape. Bev Tucker alerted his comrades after hearing the men whispering in the language of their native canton, where Tucker had gone to school. The unfortunate conspirators faced a circle of bayonets when they tried to make their break from a train en route to a prison camp in Salisbury, Maryland.

SPANDEX PAS DE DEUX

A respected businessman and community leader was charged with indecently exposing himself in three Des Moines beauty parlors after being positively identified by beauticians. J. D. Mullen, a former director of the Chamber of Commerce, was said to have entered the Xsalonce, Body Bronze, and Professional Image salons dressed in Spandex leggings, to carry out acts of "bizarre exhibitionism." The resulting publicity caused Mullen to lose his job and be shunned by neighbors.

Seven months later, however, all charges were dropped when the authorities realized they'd got the wrong man. It had been noticed that Mullen bore an amazing resemblance to Michael Long, known to police as "Spandex Man," who had been arrested many times for similar behavior. Witnesses agreed that the likeness was uncanny. Mullen said, "It's ripped my family apart. Even though the charges have been dropped, the damage has been done."

POETIC INJUSTICE

It's a humbling enough experience for any writer to find his precious book priced at a few cents on the shelf of a thrift store; worse to find it in the trash container outside a thrift store. No doubt some writers deserve such a comedown but not poet Simon Armitage, who was both humble and courageous enough to relate the tale in the book *Mortification: Writers' Stories of their Public Shame*. Armitage certainly didn't deserve the mortification that was yet to come. Taking the book out of the trash he saw that it was a signed copy. Beside the signature, in his own handwriting, were the words, "To Mum and Dad."

GIELGUD'S GAFF

The great actor Sir John Gielgud was famous for his gaffs. One night at a Hollywood party he complained loudly about a dinner party he'd attended the night before. "Terrible night," he said, "with that insufferable George Axlerod. Does anybody know him?" In the awkward silence that followed a man could be heard clearing his throat. "Well . . . I'm George Axlerod. . . ."

This begs the question: was George Axlerod's attendance at the second party really a coincidence, or was he sent there by the Devil?

FANCY MEETING
YOU HERE

It's not just a small world. It's a small world with billions of people in it. At the time of writing this sentence there were, according to the Internet World Population Clock, 6,385,725,918 people living on this planet (assuming you are reading this on Earth). That number will have grown considerably by the time you read this. Little wonder that we keep bumping into each other!

To further understand the nature of these kinds of coincidences we must also factor in the theory of "six points of separation." First imagine a very, very large field. Into this field we place all the people we know. We then add all the people that those people know, plus all the people they know, plus all the people they know, plus all the people they know, plus all the people they know.

That, according to the theory, would be all the people in the world, including Himalayan hermits and Aborigines on walkabout in the Australian outback. Just try it if you don't believe us.

Maybe the truly surprising thing is that in such a small, densely populated world, amazing chance encounters don't happen more often. Perhaps Peter Cook and Dudley Moore had it right in this comic sketch:

PETER: Hello.

DUDLEY: Hello.

PETER: How are you?

DUDLEY: I'm terribly well. How are you?

PETER: I'm terribly well as well.

DUDLEY: I must say you are looking very fit.

PETER: I'm feeling pretty fit actually. Isn't it amazing—us just bumping into each other like this?

DUDLEY: Yes. I mean here of all places.

PETER: Here of all places! I mean, I haven't seen you since, er . . .

DUDLEY: Now, er . . . hold on a minute . . . er, when was it? Er . . . we, we haven't seen each other . . .

PETER: Well actually we haven't seen each other. . . .

DUDLEY: We haven't seen each other . . . er . . . before.

PETER: That's right. We've never seen each other before, have we?

DUDLEY: No.

PETER: You've never seen me.

DUDLEY: And I've never seen you. What a small world.

PETER: What a small world!

Here's a selection of extraordinary chance encounters between people who, unlike Pete and Dud, were in fact connected.

FAR AWAY AROUND THE CORNER

Nellie Richardson said good-bye to her brother Joseph in the early 1940s and didn't see him again for more than half a century. Joseph was then a teenager, enlisting in the navy.

Nellie grew old and gave up hope of ever seeing him again, but one day, sitting in her nursing home, she was galvanized by the sight of a seventy-nine-year-old man on the other side of the room. She knew immediately it was Joe.

Perhaps as incredible as the meeting was the fact that their paths had traveled so close to each other but failed to connect for so long. At the time they met, Joe had been living in the nursing home for six months, and for decades before that the brother and sister had been living barely a mile apart in the same city.

Both had a fifty-five-year-old daughter named Sandra.

MARCIA AND PETER MEET AGAIN

Peter and Jean and Paul and Marcia are two couples who lived a couple of miles away from each other and had a mutual friend who had never introduced them. One evening the friend organized a dinner dance for eighty people and as chance would have it, Marcia and Peter were seated next to each other, for all either of them knew, for the first time.

Peter looked at her name card and said, "I'll remember your name because sixty years ago I used to play with a little girl called Marcia in India."

Marcia said, "And I used to play with a little boy called Peter."

They had both regained a childhood friend.

HE AIN'T HEAVY

You never know who you are going to meet when you hitchhike.

Tim Henderson's parents divorced when he was quite young. His father remarried and had another son, but Tim had never met him. That was until Tim hitched a lift in a car driven by diving engineer Mark Knight. During their long ride they discovered they were brothers.

O BROTHER WHERE ART THOU?

We've all had the experience of losing something and then finding it right under our noses. It happened to Rose Davies—with her brother.

Rose was just three months old when she was adopted. Years later she discovered that she had three brothers—Sid, John, and Chris—and set out to trace them.

Rose found Sid first and then John, but she didn't have to look far to find Chris.

Rose, forty-one, was staggered to discover that her long-lost brother was the man who had just moved in across the street.

"I'd only known the family for three months," said Rose. "But I thought they were very nice."

Chris, thirty-seven, was equally astonished when Rose told him who she was. He'd been searching for her, too.

HAPPY FAMILIES

Martin Plimmer and his wife were close friends with two couples; let's call them Janet and John and Antony and Cleopatra. The three couples had two children each and, come the summer, they would frequently vacation together, either the Plimmers with Janet and John, or the Plimmers with Antony and Cleopatra; sometimes all three couples would go together and the children would hit each other with plastic buckets.

This happy state of affairs changed overnight when Antony and Cleopatra split up. Antony left home and Rudolpho moved in with Cleopatra. The friendship dynamic teetered and swayed. The fundamental problem was that Janet didn't like Rudolpho. Rudolpho, in turn, didn't think much of Janet. At their first and only social meeting Janet and Rudolpho quickly got into an argument that flared into an exchange of insults and, in the next few weeks, matured into an enduring hard-boiled resentment. What's more, Cleopatra resented Janet for not liking her new love Rudolpho, and Janet resented Cleopatra for taking his side.

This chopped a third off the vacation group potential. It was generally agreed that the only thing to do was for the Plimmers to go on vacation this year with Janet and John, leaving Cleopatra and Rudolpho to explore their new-found love alone.

All this brou haha meant that booking had been put off to the last minute, and the vacation dates loomed. The Plimmers and Janet and John decided to look for a house to rent in Provence. Cleopatra and Rudolpho, working independently, decided to look for a house to rent in Provence. The Plimmers and Janet and John found a sunny house in a sleepy little village called St. Antonin du Var. Cleopatra and Rudolpho, working independently, found a sunny house in a sleepy little village called Pontevès. Then the landlord rang to apologize. He had double-booked the house, but he had another property he could let them have slightly cheaper at St. Antonin du Var, where in the course of their vacation, they all bumped into each other.

SNAP

It's odd to make friends with someone and then realize in retrospect that you've seen them before when you neither "saw" them nor knew who they were. It's odder still to have taken a photograph of them.

Graham Freer's hobby is taking and processing photographs. One of many he has taken is a shot of people in the town hall square. This particular picture was notable at the time of development only in that it was the one he chose at random to enlarge a section to test the quality of his enlarger's lens. Years later he made friends with a girl and the next time he revisited his old photographs, the previously indifferent test photo had become personal and poignant. The picture hadn't changed, but the photographer had and so had one of those distantly uninvolved girls in the square. She was his new friend.

BLOWIN' IN THE WIND

A plumber's wife had a rare visitor on the doorstep of her house in Crouch End, a quiet residential suburb of North London. It was Bob Dylan.

"Is Dave in?" he asked.

The plumber's wife explained that her husband Dave had popped out but would be back shortly. If he would like to come in and wait she'd make him a cup of tea.

The tale of Bob Dylan's unexpected call at the wrong house in Crouch End, with its delicious expectation of Dave the plumber returning a few minutes later to find the singing legend sitting in his living room drinking tea, has entered local folklore. Normally it is told and received with the kind of skeptical awe reserved for urban myths. But the story is true.

Dylan had been interested in working with the English producer Dave Stewart, founder of the Eurythmics, who has a studio in Crouch Hill. Dylan had decided to pay him a visit, but with so many road names in the area starting with the word "Crouch," he'd ended up knock-knock-knocking on the right number door in the wrong street.

.4.

LOST AND FOUND

A backpacker tells the story of losing a contact lens while bathing under a waterfall on the lower slopes of a mountain in Peru. Three days later she was washing her underwear in the river further downstream when she saw something glinting on a rock by the water's edge. It was her missing contact lens. Too amazing to be true? Possibly. But the fact remains that if something *can* happen it will, given enough time. So if you lose your contact lens in a mountain waterfall, don't give up hope. Just be patient.

And coincidence stories themselves, though not actually lost, can be found in some strangely coincidental ways. While researching this book, Brian King went into the BBC's offices to work at a computer. He was explaining to one of the department's producers, Amanda Radcliffe, that he was looking for stories for a book about coincidence when she said, "Look at the e-mail I've just received from my friend Cathy in Australia." Among some general chitchat, Cathy had written about an extraordinary coincidence she had recently witnessed. Here is the story, "Two Rings in the Bay," along with a variety of other remarkable tales of people and things lost and found.

TWO RINGS IN THE BAY

Graham Cappi of Bristol was devastated when he lost his wedding ring while on honeymoon in Nelson Bay in Australia. It fell into deep water, beyond any hope of retrieval. Graham returned to England not expecting to see the ring again.

Fifteen months later, another Englishman, Nick Deeks, was on holiday at Nelson Bay and lost his wedding ring while snorkeling. He returned the next day in the forlorn hope of finding it. After several dives he finally surfaced triumphant with a ring. But it wasn't his—it was Graham Cappi's. Encouraged by the find, he carried on diving and, incredibly, eventually found his own wedding ring.

He had no way of knowing who might be the owner of the other ring, but by chance, a few inquiries in the local town turned up some people who remembered an English visitor who had been making inquiries about a ring. Connections were made and Graham Cappi was eventually contacted in England. He was overjoyed to learn that his wedding ring had been found. It was returned to him by a young local girl making a planned trip to England. She was intrigued to discover that Graham's wedding date, inscribed on the ring, was also the date of her birthday.

THE GOLDEN MATCHBOX

The golden matchbox was a gift from the Prince of Wales to his friend and fellow fox hunter, Edward H. Sothern—a successful actor in the 1890s.

Out on a hunt one day, Sothern fell from his horse and the matchbox broke from its chain and was lost. Sothern had a duplicate made, which, after his death, went to his son Sam. Sam, also an actor, took the matchbox on a trip to Australia where he gave it to a Mr. Labertouche. Returning to England, Sam learned that the original matchbox had been retrieved, twenty years after its loss, by a farmhand who had found it while ploughing a field that very morning.

Sam explained what had happened in a letter to his brother Edward H., the third actor in the family, who was touring in America. Edward read it while traveling on a train with a companion, Arthur Lawrence. Edward told Lawrence the story, which prompted his friend to take a watch chain from his pocket. Dangling from its end was the duplicate golden matchbox—a gift from the Australian Mr. Labertouche.

WHERE THERE'S MUCK

Barbara Hutton accidentally flushed her antique bracelet down the toilet. Months later she was in a jeweller's when a man brought in a bracelet to be valued. It was Barbara's. The man had found it while working in a sewer.

BACK FROM THE DEAD

Alpha Mohammed Bah feared his partner and children were dead. He describes the moment when coincidence reunited them as "like being born again."

In early 1997 Alpha was working as a commercial photographer in Freetown, the capital of Sierra Leone. His partner Fatmata and their two daughters Sordoh and Marian lived across the city.

"It was a good life—I was self-reliant, making a living for my family," he said.

But everything changed when a military junta seized power and Alpha was forced to become the head of security for his local community.

"Everyone was very afraid of the government," he said. "Random executions and other atrocities were happening day to day. I couldn't bring myself to prosecute innocent people—but if I refused I would be shot—so I decided I would have to leave."

Alpha was forced to flee without saying good-bye to Fatmata and his daughters—aged three and a few months at the time—as they lived on the other side of Freetown and he could not cross the checkpoints. He fled to New Guinea and waited for the situation to change back home.

In 1998, the junta were overthrown for a short time, and Alpha returned to try to trace his family. "I searched in many displacement camps for my relatives, but could not find them," he said.

"During the year I was there I did not meet anyone who could tell me whether they were alive or dead. I started to believe they were dead because I had nothing to keep my hope alive."

As the political situation began to deteriorate again, Alpha made the decision to emigrate to New York. En route, he was detained and questioned by immigration officials at Heathrow. He decided to seek asylum in Britain.

He settled in Wales, where he helped other refugees with translation and completing their asylum application forms. Some months later, he was contacted by a friend who asked him if he could help a woman and two children from Sierra Leone who had just arrived in Britain.

To his amazement he found it was Fatmata and his daughters.

"I could not believe it," he says. "She just started weeping—and I was crying tears of joy. She later told me she had given up hope of ever seeing me alive again."

HOMING BRUSH

During the First World War, a U.S. soldier was on board a troop ship torpedoed off the French coast. He survived, but lost all his possessions. He also survived the rest of the war. In America after the war he was by the seashore near Brooklyn when he found a shaving-brush cast up on the shore. It had an Army number on the back. It was his own shaving brush.

A RADIO YOU CAN RELY ON

Canadian Mike Mandel tells this story about his father, who was active in the amateur radio scene in Toronto in 1976. Mike's father knew his subject, having been a radio communications tutor with the British army during World War II.

One evening a radio enthusiast friend came round to pick up some gear he had bought from Mike's father and the pair lugged it from the basement of the townhouse to the car outside, where they stood chatting.

Mike's father asked the friend if he had ever heard of a 19 set. The friend shook his head and Mr. Mandel explained that the 19 set was an old Russian tank radio that he had used to train radio operators during the war. The gauges were in Russian but the radio had the advantage of being reliable and simple to use. He hadn't seen one since the war, he said, but he'd love to get hold of one.

They said good-bye and Mike's father was walking back to the house when he noticed a pile of junk at the bottom of a wall. On top of the pile of junk was a 19 set Russian tank radio.

"Dad said he got a chill when he saw it," said Mike. "He took it from the junk pile and carried it to our townhouse. He plugged it in and it worked perfectly. All the ancient tubes were intact."

They made enquiries but never discovered where it came from. When Mike's father died the 19 set was donated to Ottawa's National War Museum, which displayed it with a photograph of Mr. Mandel senior seated at it.

PHOTO FIT

Colin Eves was standing outside his local shopping mall when a man approached him and introduced himself as Derek from the local post office.

"I've seen a photo of you," he said. "You were looking out over the harbor." Derek took Colin back with him to the post office and produced some prints that had been found loose in some incoming mail. They were indeed pictures of Colin. His mother had taken them on a visit to see her son and had sent them to him but the envelope split open during processing at the post office.

A WARM WELCOME FOR JACK FROST

Novelist Anne Parrish was excited to find a copy of *Jack Frost and Other Stories*, published in English, in one of the secondhand bookstalls beside the Ile de la Cité, in Paris. It had been a favorite book in her nursery in Colorado Springs, but she had not seen a copy since she was a child. She showed the book to her husband, who opened it at the title page, where he found the inscription: "Anne Parrish, 209 N. Weber Street. Colorado Springs."

A DIARY'S SECRET ENTRY

A diary lost in a field in 1952 turned up just over a year later at the feet of its owner, who had stopped to light a cigar in the same field. Leon Goosens, a famous oboist, picked up the battered object and flicked through it. The bindings had sprung apart and inside he could see that the covers had been stiffened with squares of newspaper. This was normal practice in book binding at the time, so there was nothing unexpected about that, but what gave him pause was that this particular scrap of newspaper was about him. It was a piece from a nineteen-year-old gossip column about his marriage in 1933.

MEET THE FAMILY

Sometimes things we have lost turn up in the most unexpected circumstances, many years after they disappeared. In the case of Kari Maracic, it happened to be her brother.

Ila Manner was only seventeen when she discovered she was pregnant by a young surfer named Chris Maracic she had met at a Florida high school dance. Maracic had just been drafted to Vietnam so when the child, a boy, was born, Ila's parents arranged for him to be adopted. On the adoption form Ila wrote that the parents' occupations were hairstylist and oceanographer, jobs they had dreamed of doing.

Eventually Maracic returned from Vietnam and the two were married. A year later, they had a daughter they named Kari.

The son, named Ben, grew up knowing he was adopted, but not who his real parents were or that he had a sister. "While I was in high school I took a test to determine what I was going to be when I grew up," Ben said. "One of the options that I was given was an oceanographer. My foster dad said that was amazing because the adoption papers had said that my real father was an oceanographer."

Kari meanwhile had moved to San Francisco where she shared a room with a girl named Erin Kehoe. One evening Erin invited Kari to a dinner. Also at the meal was a friend of hers named Ben Davis. At some point in the evening Erin asked Kari about her long lost brother.

Kari told how she had looked for her brother for eleven years without luck. Ben told her that he had been adopted from Florida. This in itself seemed quite a coincidence. Kari then told him the date of her brother's birthday and a startled Ben replied, "That's my birthday." Erin looked hard at them both. "Hey you guys do look kind of alike," she said.

But the possibility that they were brother and sister seemed to be dashed when Ben told them the occupations his natural parents had put on the adoption papers.

When Kari next spoke to her father on the phone she told him about the meeting. He became animated when she mentioned the adoption paper occupations of Ben's parents. He told her that at the time Ila had been studying to be a hair stylist and he had planned to be an oceanographer. At this point, said Kari, her stomach turned over. She knew she had found her long lost brother.

VIETNAM JACKET

The anonymous winner of a "strange but true" story competition posted the following account on the Internet.

"One weekend I went with a new male friend to a local flea market. My pal—a Vietnam veteran—had voiced a casual interest in finding a fatigue field jacket and I was keeping my eyes open for one.

"I spotted a field jacket and for some reason looked inside the cuff, where I noticed my friend's last name printed in black marker. It's a somewhat unusual name, and my first thought was that this was strange, and I wondered first if the seller had heard us talking and was pulling a gag, but that was ridiculous.

"I just said, 'Look at this. Isn't that weird,' and then I looked up at his reaction and his face had gone white. He didn't speak, he just gulped and nodded.

"It was his jacket from Vietnam. He had turned it in to the army when he got out. He bought the jacket and wears it occasionally."

5

LIFE IMITATING ART

The scene is familiar to millions of moviegoers. Leonardo DiCaprio as Jack Dawson and Kate Winslet as Rose DeWitt Bukater plunging into the icy waters of the North Atlantic after the mighty ocean liner the *Titanic* sinks below the waves to the wails of the drowning and Celine Dion.

As a historical record, James Cameron's blockbuster leaves a lot to be desired, but that's beside the point here.

The point is that this celluloid reconstruction of the tragic sinking of the RMS *Titanic* was an example of art imitating life. What many people don't know is that the real-life events surrounding the *Titanic's* maiden voyage in 1912 appears to be an example of life imitating art.

The *Titanic* is by no means the only example of this eerie phenomenon.

THE *TITAN* AND THE *TITANIC*

No one has ever come up with a theory that even begins to explain the extraordinary parallels between the novella *The Wreck of the Titan or, Futility*, written by American writer Morgan Robertson in 1898, and the real-life events surrounding the sinking of the RMS *Titanic* some fourteen years later in 1912.

Robertson might as easily have been writing a piece of journalism describing the tragic sinking of the *Titanic*, so similar were many of the details. The month of the wreck, the number of passengers and crew, the number of lifeboats, the tonnage, length, and even speed of impact with the iceberg were all close to identical.

The novella is worth looking at in some detail, so remarkable are the similarities with the tragedy it so uncannily foretells. It begins:

> *She was the largest craft afloat and the greatest of the works of men. In her construction and maintenance were involved every science, profession, and trade known to civilization. On her bridge were officers, who, besides being the pick of the Royal Navy, had passed rigid examinations in all studies that pertained to the winds, tides, currents, and geography of the sea; they were not only seamen, but scientists.*
>
> *. . . Two brass bands, two orchestras, and a theatrical company entertained the passengers during waking hours.*
>
> *. . . From the bridge, engine-room, and a dozen places on her deck the ninety-two doors of nineteen water-tight compartments could be closed in half a minute by turning a lever. These doors would also close automatically in the presence of water. With nine compartments flooded the ship would still float, and as no known accident of the sea could possibly fill this many, the steamship Titan was considered practically unsinkable.*
>
> *. . . She was eight hundred feet long, of seventy thousand tons displacement, seventy-five thousand horse-power, and on her trial trip had steamed at a rate of twenty-five knots an hour over the bottom, in the face of unconsidered winds, tides, and currents. In short, she was a floating city—containing*

within her steel walls all that tends to minimize the dangers and discomforts of the Atlantic voyage—all that makes life enjoyable.

Unsinkable—indestructible, she carried as few boats as would satisfy the laws. These, twenty-four in number, were securely covered and lashed down to their chocks on the upper deck, and if launched would hold five hundred people.

The simple statistics of the comparisons between Morgan Robertson's *Titan* and the *Titanic* are remarkable. But it's Robertson's description of the *Titan's* collision with the iceberg that is so chillingly prescient of the real-life events fourteen years later.

Two bells were struck and answered; then three, and the boatswain and his men were lighting up for a final smoke, when there rang out overhead a startling cry from the crow's nest.

. . . "Ice," yelled the lookout. "Ice ahead. Iceberg. Right under the bows." The first officer ran amid-ships, and the captain, who had remained there, sprang to the engine-room telegraph, and this time the lever was turned. But in five seconds the bow of the Titan began to lift, and ahead, and on either hand, could be seen, through the fog, a field of ice, which arose in an incline to a hundred feet high in her track.

. . . seventy five thousand tons—dead-weight—rushing through the fog at the rate of fifty feet a second, had hurled itself at an iceberg.

. . . The holding-down bolts of twelve boilers and three triple-expansion engines, unintended to hold such weights from a perpendicular flooring, snapped, and down through a maze of ladders, gratings, and fore-and-aft bulkheads came these giant masses of steel and iron, puncturing the sides of the ship, even where backed by solid, resisting ice; and filling the engine and boiler-rooms with scalding steam, which brought a quick, though tortured death, to each of the hundred men on duty in the engineer's department.

Amid the roar of escaping steam, and the bee-like buzzing of nearly three thousand human voices, raised in agonized screams and callings from within the enclosing walls, and the whistling of air through hundreds of open dead-lights as the water, entering the holes of the crushed and riven starboard side, expelled it, the Titan moved slowly backward and launched herself into the sea, where she floated low on her side—a dying monster, groaning with her death-wound.

Late on the night of Sunday April 14, fourteen years after *The Wreck of the Titan* had been published, the RMS *Titanic,* heralded as "practically unsinkable" by its owners, the White Star shipping line, struck an iceberg and was holed below the waterline. Less than three hours later she had sunk beneath the waves. Of the 2,200 aboard, only 705 people, mainly women and children, were rescued.

BREAK A LEG

Shortly before the opening night of the musical *42nd Street,* entertainer Jan Adele slipped and tore the ligaments in her left ankle. The plot of the show concerns a Broadway director looking for one more hit before he retires. His hopes are dashed when his leading lady twists her ankle just before opening night.

SNAKE

A deadly poisonous viper came within inches of killing actress Trudie Styler—also the wife of Sting—in the Brazilian rainforest, but by coincidence she knew how to react because she had just finished making the film *Fair Game,* in which she plays a woman locked in a flat with a killer mamba. She believes the film saved her life.

The coincidence is all the more remarkable for the fact that Styler is phobic about snakes, and never intended to be in the film at all. "I can't look at a picture of a snake. If one comes on the television screen I leave the room; it's that kind of reaction."

The odds were against her taking the part. When her agent suggested it, she said, "Forget it!" The Italian director Mario Orfini didn't seem interested in her either.

"He said, 'Very nice to meet you, but I'm looking for Kim Basinger,' and I said, 'Well you'll be looking for a long time because you haven't got the kind of money that she wants.'"

"Then he said, 'If she won't do it the next person on our list is Rosanna Arquette.' I said, 'She's a friend of mine, she's doing movies back to back for the next year, but I look a bit like her.' I had no intention, if he offered it to me, of accepting, but my pride wouldn't let me back down."

Eventually an offer was made to Styler and she accepted, mainly because she realized how unusually prominent the female lead role was.

Four months later, Styler was in the middle of the tropical jungle in Brazil with Sting, while setting up their Rainforest Foundation charity. One night she woke up with a presentiment that something wasn't right. "I got out of my hammock, feeling deeply uncomfortable, put my feet on the ground (I had naked feet), got my flashlight and walked a couple of feet, then my body froze as if to say, don't go any farther. I shone my flashlight and there, reared up in front of me, was this big snake, mouth open, ready to strike. My hand was six inches from his face, and he could have got me if he wanted, but I kept very still. I only knew that because of the movie, because this was an exact reenactment of a scene in that film.

"I breathed very slowly and deeply because snakes can sense panic, and they're deaf, so I knew if I kept rigid I could shout for help. I said, 'Sting, there's a snake!' And the wretched man said 'Huh?' It's very irritating when someone doesn't hear you and you're just about to be killed! So I said it a bit louder and this time he did hear and also the natives must have heard because they woke up and came into the tent with a club and killed it."

THE *CAROLINE*'S DOUBLE DISASTER

Playwright Arthur Law was astonished to find that a play he wrote appeared to predict an actual event. His play, *Caroline*, written in 1885, was about Robert Golding, sole survivor of a wrecked ship called the *Caroline*. Just days after the play opened, a ship, the *Caroline*, went down. Its one survivor was a man named Golding.

THE MAN WHO INVENTED HIS WIFE

An idealized girl created for a novel came intoxicatingly to life in a Berlin café in 1929, when German playwright and novelist Leonhard Frank believed himself to be looking at the very girl who had haunted his book and imagination. Frozen by careering emotions and conscious of his age (he was forty-eight; she must have been no more than twenty), Frank watched her for too long without addressing her. A youth came into the café, apologized to her for being late, and swept her out of his life.

Frank haunted the Romanisches Café for weeks afterward, hoping she would reappear, but she never did. He had to wait another nineteen years for the chance to talk to her.

Three years passed and Frank was forced to leave Germany to avoid persecution by the Nazis.

He had written the book, *The Singers*, in 1927. In it he had invented a character, Hanna, who stood for all the best qualities of young womanhood as he saw them. She was graceful, slender, hot blooded, with an olive and rose-colored complexion, and she projected emotional strength, humor, high spirits, and an irresistible curiosity about life. His sighting of the living Hanna in the Romanisches Café occurred two years later, while recovering from the effort of writing a further novel. He couldn't stop looking at her. She was everything he had dreamed she would be. But she didn't see him.

In the summer heat wave of 1948 Frank was in America, where he had found work as a Hollywood scriptwriter. He was living in New York but had escaped the heat and fled to a farm in the countryside that took paying guests. It was there he saw "Hanna" again, sitting just as he remembered her. He spent a day collecting his senses and then approached her. He told her about first seeing her in Berlin, about how she resembled the idealized girl in his book ("Hanna," she said, encouragingly) and then tried to kiss her. She rejected him. She was married to the young man he had seen at the café, she said. Her name was Charlotte.

She avoided him for three weeks, but they met again and this time his feelings for her were reciprocated. The next morning she rang her husband to ask for a divorce. At the end of a long chain of extraordinary events, Frank married his Hanna.

For Frank the story "confirmed once again my belief that accident in human life may be synonymous with destiny."

A CASE OF ART IMITATING LIFE

A wardrobe department buyer visited a Los Angeles second-hand clothing shop to find worn yet elegant clothing for Frank Morgan's character, Professor Marvel, in the MGM screen adaptation of L. Frank Baum's *The Wizard of Oz* (1939). They returned with a pile of suitable coats. After trying several on, a well-worn Prince Albert coat of black broadcloth, with a splendid velvet collar was chosen. It fitted Morgan perfectly. Later, after filming had begun, Morgan was inspecting the coat in detail and was astonished to find the name "L. Frank Baum" stitched inside.

SPY CENTRAL

Norman Mailer wrote *Barbary Shore*, his novel about a writer and a group of spies, while living at 102 Pierrepont Street, Brooklyn, New York. At first he had no intention of writing about spies, but as he got going on the book he introduced a Russian spy and gradually the character started to dominate. After the book was published in 1951, the U.S. Immigration Service arrested a man who lived on the next floor to Mailer. He was Colonel Rudolf Abel, the most wanted

Russian spy in the States at that time. Playwright Arthur Miller lived in the same building, too, though Mailer didn't write a book about playwrights.

THE TWIN TOWERS COVER VERSION

In what appears to be an extraordinary example of prescience, the image of the devastating attack on New York's Twin Towers was anticipated on the planned cover of a hip-hop album due to be released just weeks after the September 11 event.

The cover of the *Party Music* album by the hip-hop group The Coup depicted the band with an exploding World Trade Center in the background. A group member waves two sticks held between thumbs and forefingers as if "conducting" the proceedings.

Band member "Boots" Riley explained the symbolism behind the cover, "I came up with the idea with the photographer. We took the pictures on May 15, and we were done with it by the beginning of June. Any similarities are totally coincidental, and it was originally supposed to be more of a metaphor for destroying capitalism— where the music is making capitalist towers blow up."

The album was due for release in November 2001. The cover artwork was rapidly redesigned.

DEADLY WORDS

Columbian drug baron Pablo Escobar read of his own death. Tom Clancy, the author of *Clear and Present Danger*—which later became a Hollywood box-office success—based his fictitious drug baron on Escobar. Clancy describes how his drug baron is shot dead by the Columbian national police as a result of an intercepted cell phone call he makes to his family. In real life the police used a computer that identified Escobar's voice on the phone and within minutes located him and moved in for the kill. A heavily annotated copy of Clancy's novel was later found in Escobar's apartment, with the scene relating to the phone call underlined. On the day Escobar was killed, the same scene was being filmed.

THE CABIN BOY—RICHARD PARKER

In 1884, seventeen-year-old Richard Parker ran away to sea and became a cabin boy aboard the ship the *Mignonette*. The rest of the crew consisted of the captain Thomas Dudley, mate Edwin Stephens, and hand Edmund Brooks. They left Southampton bound for Australia.

They were 1,600 miles from land when a South Atlantic hurricane broke. The *Mignonette* was hit by huge waves and sank. In the panic to board the lifeboats, the crew were unable to salvage any provisions or water except two small tins of turnips.

The crew had very little to eat or drink for nineteen days and became desperate. Richard Parker drank sea water and became delirious. Captain Dudley considered drawing lots to choose a victim to feed the remaining crew. Brooks was against any killing whatsoever, and Stephens was indecisive so the Captain decided to kill the boy as he was near to death and had no dependants.

They said some prayers over Richard's sleeping body. Dudley then shook him by the shoulder and said, "Richard my boy, your time has come." The three sailors dined and survived on Richard's carcass for thirty-five days until rescued by the aptly named vessel SS *Montezuma*—named after the cannibal king of the Aztecs.

The resulting court case fascinated Victorian society and became the best-documented study of cannibalism in the UK. Dudley, Stephens, and Brooks were each sentenced to six months' hard labor and later emigrated.

But the story has a strange twist in its tail. Half a century before the grisly events, in 1837, Edgar Allan Poe wrote *The Narrative of Arthur Gordon Pym of Nantucket*. This book tells of four shipwrecked men who, after many days' privation, drew lots to decide who should be killed and eaten.

The cabin boy drew the short straw. His name was Richard Parker!

RICHARD PARKER POSTSCRIPT

It seems that coincidences beget coincidences.

Craig Hamilton Parker's grandfather was a cousin of the young cabin boy Richard Parker. Craig has recorded a whole string of further coincidences connected with his ancestor's tragic story.

"My cousin Nigel Parker was the first to notice the link between the Poe story and actual events. He wrote an account and sent it to Arthur Koestler who published it in *The Sunday Times* on May 5, 1974.

"Koestler, author of *The Roots of Coincidence*, relates how sometime after the news story, he casually mentioned it to John Beloff at the University of Edinburgh, who had, that day, written about it in his journal.

"Nigel's father, Keith, thought that Richard's story would make an interesting theme for a radio play and began to plan a synopsis. At that time, to supplement his writer's income, he reviewed books for Macmillan publishers. The first book to arrive through the post was *The Sinking of the Mignonette*. A few weeks later he was asked to review another play, among a collection of short plays, called *The Raft*. It was a comedy for children with nothing sinister about it at all, apart from the cover illustration. Three men seemed to threaten a young boy, which is completely out of keeping with the play's tone. *The Raft* was written by someone called Richard Parker.

"In the summer of 1993, my parents took in three Spanish-language students. My father told them about Richard Parker one evening over supper. The television was on in the background. All conversation stopped when a local program started talking about the remarkable story. Dad went on to break the silence by saying how weird coincidences always occur whenever Richard's tale is mentioned. He told them about Edgar Allan Poe.

"Two of the girls went white. 'Look what I bought today,' said one. She reached into her bag and pulled out a copy of the Poe story. 'So have I,' said the other girl. Both had gone shopping that day and independently bought the very same book containing the Richard Parker story."

PATTY HEARST

Patty Hearst, daughter of a wealthy media tycoon, was kidnapped in 1974 by a radical terrorist group called the Symbionese Liberation Army. It was one of the strangest kidnapping cases ever.

Even stranger was the fact that a pornographic novel called *Black Abductor* by James Rusk Jr. (writing under the pseudonym Harrison James) had been published two years earlier, which described many of the facts of the Hearst story with startling accuracy.

As soon as the kidnapping made world headlines, the publishers, Dell-Grove, wasted no time reissuing the book with a new title, *Abduction: Fiction Before Fact*. Though often shrouded by lurid accounts of sex, the story contains many eerie parallels to the Hearst case.

It tells of a young college student named Patricia, daughter of a wealthy and prominent right-wing figure, who is kidnapped near her college campus while walking with her boyfriend.

The boyfriend is severely beaten and becomes a prime FBI suspect before being cleared.

The kidnappers belong to a multiracial group of radical activists who model themselves on Latin American terrorists and are led by an angry young black ex-convict.

They demand the release from jail of a comrade who is imprisoned for a political assassination, sending Polaroid pictures of the girl with their communications to her father and describing the abduction as the United States' first "political kidnapping."

At first the girl is an unwilling captive but later she becomes receptive to the group's aims and joins them.

The fictional abductors predict that eventually their hideout will be found by police and they will be surrounded, teargassed and killed.

Given such similarities, the FBI were bound to consider whether Rusk had been in on the planning of the kidnapping or the SLA had got the idea from reading his book.

MARIE COLLIER

On December 8, 1971, opera star Marie Collier fell to her death from the balcony of her home. She had been talking to her financial adviser about a new tour when she opened a window and fell out. Marie Collier had come to fame in the role of Tosca, who leaps to her death from atop a wall in the last act. It was Collier's last role before she fell from the window.

6

TIMELY
MANIFESTATIONS

You've worked yourself into a fine lather, addressing your coworkers on the gross inadequacies of the boss—his arrogance, his general incompetence, his failure to grasp even the barest essentials of management. Suddenly, it dawns on you that you have lost the attention of your audience. They are all staring at a point just behind you. You turn to discover the great man standing in the doorway—glaring at you.

Such manifestations are the downsides of coincidence. Sometimes, thankfully, they can work in our favor.

THE CONQUEST OF SPACE

Charles Carson had a frustrating problem. He was making slides of book illustrations to show at a talk he was about to give to his local astronomical society. But he was missing the most important book and he couldn't find it anywhere.

The talk was about artists who paint representations of outer space. Carson had lots of paintings of outer space, but none by the most respected artist in the field, Chesley Bonestell. The book he needed, *The Conquest of Space*, was full of Bonestell's paintings, but a search on the computer at his local library revealed that there wasn't a single copy anywhere in town.

A couple of hours later his wife went into a charity shop and he followed aimlessly. Of course, he looked at the books on sale, but there was nothing there to interest him. Stuck for something to do to pass the time, he began idly leafing through a row of children's books. Among them was *The Conquest of Space* by Chesley Bonestell.

It's a Small Island

Before he became a bestselling author, Bill Bryson worked as a freelance journalist. At one point a magazine commissioned him to write an article on remarkable coincidences. He managed to gather a lot of information but didn't have enough examples to fill out the piece, so he wrote to the magazine to say he couldn't do it. He put the letter aside to mail the next day. Going into the *Times* the next day he noticed an announcement of a sale of books that had been sent in for review. The first book that caught his eye was *Remarkable True Coincidences*. He opened it at a story about a man named Bryson (see page 193). The book had found its man.

The Actor and the Book

In 1971 George Feifer's personal copy of his novel, *The Girl from Petrovka*, heavily annotated in the margins, had been stolen from his car in a London street. Two years later the film rights of the novel were sold and Anthony Hopkins was cast to play the lead.

Hopkins tried to buy a copy of the book, but despite trying several bookshops in London, there was not a copy to be had. Disappointed, he started his journey back home. On the way he noticed an open parcel on a seat in London's Leicester Square Underground station. He half suspected a bomb and inspected it with caution, but it was a book—George Feifer's *The Girl from Petrovka*—the very book he had been trying to buy. Later, meeting Feifer in Vienna, Hopkins showed the author the book. It was the author's personal copy, stolen two years previously.

ALWAYS GO TO THE TOP
(UNLESS THE TOP COMES TO YOU)

Photographer's agent Mark George celebrated his father's eightieth birthday by buying him dinner at the Grill, in London's Savoy Hotel. During the course of the meal, he told the story of a woeful experience he'd had with a ill-tempered manager in a hotel at Lochinver in western Scotland.

Mark had been on a diving vacation with friends and that particular evening they'd decided to have dinner at the Inver Lodge Hotel. A notable feature of the hotel's dining room is a long picture window affording spectacular views across the loch and its coastline. The restaurant was almost empty that evening but the waiter showed them to a table at the back of the room. Mark asked if they could have a window table instead. The waiter looked doubtful, and when Mark and his companions started to walk across the room, the manager, who was himself eating at a window table, stood up and stopped them, saying that the window tables were unavailable because they were being laid out for breakfast.

Bristling, they nevertheless accepted the inferior table. Halfway through the evening another friend who had been driving with them joined them for a drink. He wasn't hungry but ordered some wine. No sooner had the waiter gone than the manager reappeared, saying the friend couldn't have a drink unless he ordered food as well. Exasperated, they asked for a menu so they could order a token item of food. At this the manager told them the kitchen was closed.

"I couldn't believe anyone could be so difficult," Mark told his father, shaking his head.

At this point a man who had been eating at a table adjoining theirs at the Savoy came over and introduced himself as Lord Vestey, an English aristocrat and businessman. Vestey said he wasn't given to eavesdropping, but he couldn't help overhearing their conversation. "I'm afraid to say I own that hotel," he said. He apologized, promised to look into the matter, and invited Mark to visit the hotel again free of charge.

ICELANDIC CHESS

Arthur Koestler described the following incident as a "trivial but typical case of a frequently recurring pattern."

He wrote, "In the spring of 1972, the *Sunday Times* invited me to write about the chess championship match between Boris Spassky and Robert Fischer, which was to take place later that year in Reykjavik, Iceland. Chess has been a hobby since my student days, but I felt the need to catch up on recent developments; and also to learn something about Iceland, where I had only spent some hours in transit on a transatlantic flight during the war. So one day in May, I went to the library to take home some books on these two unrelated subjects. I hesitated for a moment, whether to go to the "C" for chess section first, or to the "I" for Iceland section, but chose the former because it was nearer. There were about twenty to thirty books on chess on the shelves, and the first that caught my eye was a bulky volume with the title, *Chess in Iceland and in Icelandic Literature*."

THE WEDDING SINGER

Tony Mills planned to ask his good friend Harriet to sing at his wedding in June 1996. He had casually mentioned it to her some months before the wedding day. As the time approached, for various reasons the reception was arranged in something of a hurry. The manager of the restaurant in which it was to be held said he would take care of the entertainment—he knew of this great singing duo. Tony realized he must phone Harriet and explain so she would not be offended. Harriet responded with the news that she had already been booked that evening to sing at a wedding reception—for Tony and his bride.

HERE'S YOUR STUPID SHOVEL

On December 23, 1946, seven-year-old Bill McCready, together with his parents and baby brother, set off in a blizzard in their old car for Christmas with their family. The journey ended in disaster when they careered into a ditch. Fortunately, they were rescued after a couple of hours by a truck driver who dug them out. He then drove off, leaving his shovel behind. Bill's father always kept the shovel in his truck as a lucky talisman. Seventeen years later, Bill, now aged twenty-four, found himself in a restaurant with his father.

Bill says, "The two of us were sitting there and two tables away three men were having lunch. One of them was talking about the bitter winter of '46 and how he had come across a family stuck in a car in a ditch. He was explaining how he had dug them out, but they had then driven away taking his "stupid shovel" with them.

"My father said nothing but got up, went outside, and came back with the shovel. He tapped the man on the shoulder and said, 'Here's your stupid shovel.'"

REQUIEM FROM A BLACKBIRD

Roy Smith's brother died in 1993. He had always had a keen interest in birds and in the time before his death he used to attract a particular blackbird into his back garden to feed from his hand. The day of his burial was marked by heavy, almost torrential rain. It was dramatically moving anyway, but there was a sweeter poignancy in store.

"We were all standing around the grave under umbrellas as the vicar read the dedication," says Smith, "when we were suddenly all aware of the sweetest birdsong. On the roof of a nearby gardener's building was a blackbird singing its heart out with a continuous, melodious song, drenched in the downpour. We were all mesmerized."

It was so rare to see a bird out in such weather, particularly singing as this one was singing, that the vicar declared he no longer believed in coincidence. "For me," said Smith, "It was the hand of God bringing comfort to us all in our grief."

TIME, GENTLEMEN

Pope Paul VI's alarm clock went off at 9:40 p.m. on August 6, 1978, but the pope didn't wake up. In fact he died at that precise moment. Even stranger, the alarm was actually set for six the following morning.

A similar event is said to have happened when King Louis XIV of France died at 7:45 a.m. on September 1, 1715, though in this case an ornate clock belonging to him is supposed to have stopped. Considering Louis XIV probably owned ten thousand ornate clocks, maybe this wasn't so remarkable.

CUNNING STUNT

The Danish tenor Lauritz Melchior was famous for his heroic Wagnerian roles with the New York City Metropolitan Opera, but the most romantic scene of his life took place when he was a student living in a small pension in Munich. He was sitting in the garden learning a part for an opera. As he sang the words, "Come to me, my love, on the wings of light," a female parachutist landed at his feet. It was the Bavarian actress Maria Hacker, who was performing a stunt for a movie thriller. They were married in 1925. "I thought that she came to me from heaven," said Melchior.

7

JINXES AND CURSES

Psychologist Professor Christopher French heads the Anomalistic Psychology Research Unit at Goldsmiths College in London. Its objective is to explore the facts behind claims of parapsychological phenomena—from ESP to alien abduction.

Professor French is skeptical that such claims have any foundation in scientific reality. He extends that skepticism to the realm of jinxes and curses—often cited as the cause of protracted runs of bad luck.

Of the Superman curse, outlined earlier in this book, he dismisses the catalog of misfortune that has befallen many people associated with the story of the Man of Steel as pure coincidence. The original creators of the Superman story sold the rights for a pittance, and many of the stars of the television and film adaptations have suffered tragic accidents and illnesses—Christopher Reeve most prominent among them. But Professor French points out that a great many people have done very well out of the whole Man of Steel business.

"I'd be very happy to be given the rights of the Superman story," he says. "I'll take my chances with the curse."

He's equally dismissive of the jinx that is said to plague those who desecrate the tombs of the pharaohs.

It is reported, for example, that in the 1890s Professor S. Resden opened an Egyptian tomb that bore the inscription: "Whosover desecrates the tomb of Prince Sennar will be overtaken by the sands and destroyed." Resden knew he was doomed, it was said. He left Egypt by ship and died on board, a victim of suffocation, from no discernible cause. Small amounts of sand were said to have been found clutched in his hands.

Professor French believes most such stories of mysterious premature deaths are nonsense. But those less skeptical than him remain open minded about what might lie behind such coincidences. So what supernatural force, curse, jinx, or darned bad luck could account for the following tales of disasters, death, and multiple lightning strikes?

THE SPORTS ILLUSTRATED COVER CURSE

Appearing on the front cover of *Sports Illustrated* ought to have been a blessing, but for hundreds of sports stars it became a curse.

The list of season-ending injuries, fatal car crashes, family tragedies, divorces, batting slumps, and losing streaks suffered by individuals and teams featured on the cover is long and puzzling.

The curse began back in 1954 when, a week after appearing on the cover of *Sports Illustrated*, Major League Baseball player Eddie Mathews suffered a hand injury that forced him to sit out seven games. It wasn't long before sporting professionals, sports journalists, and readers were making more connections between incidents of misfortune and appearances on the front cover of the magazine. Other notable cover coincidences include:

- January 31, 1955—In the week her picture appeared on the front cover, skier Jill Kinmont struck a tree during a practice run and was paralyzed from the neck down.
- May 26, 1958—*Sport Illustrated*'s 1958 Indianapolis 500 preview issue featured Pat O'Connor, who was killed in a fifteen-car pileup during the first lap of the race.
- February 13, 1961—Laurence Owen was billed as "America's Most Exciting Girl Skater." Two days after appearing on the cover, Owen and the rest of the United States figure skating team died in a plane crash.
- December 14, 1970—The University of Texas, 10–0 and enjoying a thirty-game winning streak, fumbled nine times in its next game, sustaining a 24–11 loss to Notre Dame in the Cotton Bowl.
- September 4, 1989—Major League Baseball Commissioner Bart Giamatti died of a heart attack the week after he was quoted on the cover of *Sports Illustrated.*
- June 5, 1995—Three days after his cover appearance, San Francisco Giants third baseman Matt Williams, the National League leader in home runs, fouled a pitch off his right foot, breaking it, sidelining him for nearly three months.

Sports Illustrated conducted its own investigation into the alleged curse and found that in 2,456 editions there had been 913 examples of "jinxes" in the form of a substantial misfortune or decline in performance involving an athlete featured on the cover. That represented a 37.2 percent rate of misfortune for cover stars. The investigation also revealed that the jinx affected some kinds of sportspeople more than others. Golfers, for example, were "jinxed" almost 70 percent of the time. Tennis players suffered misfortune in 50 percent of cases. But boxers seem comparatively immune, suffering bad luck after a mere 16 percent of appearances on the front cover of the magazine.

THE CURSE OF THE BAMBINO

In 1918 the Boston Red Sox became the most successful baseball team of all time when they won their fifth World Series. One of the stars of the team was a young pitcher by the name of George Herman Ruth—also know as Babe Ruth, or The Bambino.

But two years later, on January 3, 1920, Red Sox owner Harry Frazee made what appears to have been a catastrophic mistake. He sold Ruth to the New York Yankees for $125,000 in cash and a $300,000 loan, so he could finance a play called *No, No, Nanette*.

The Yankees, who had never won a World Championship before acquiring Ruth, went on to win twenty-six, becoming one of the greatest success stories in the history of sport. Meanwhile, the Boston Red Sox appeared in only four World Series after 1918, losing each one in game seven. Many consider Boston's poor performances after the departure of Babe Ruth to be attributable to "The Curse of the Bambino."

The story has a happy ending. In 2004, after eighty-six years in the wilderness, the Boston Red Sox finally became World Series Champions again, beating the St. Louis Cardinals. The god of baseball, it seems, had finally lifted the curse.

THE MUMMY'S CURSE

The Fifth Earl of Carnarvon and archaeologist Howard Carter discovered the tomb of the boy Pharaoh Tutankhamun on November 26, 1923, after years of searching.

Lord Carnarvon did not have long to enjoy his fame. In fact he didn't live long enough to even set eyes on the fabulous treasures hidden within the tomb. Just four months after finding the hidden entrance, he died from blood poisoning caused by an infected mosquito bite. He was fifty-three.

It is said that at the time of his death, lights went out all over

Cairo. The local power company could not explain it. Some reports also claim that at precisely the same moment, Lord Carnarvon's dog, back in England, suddenly howled and dropped down dead.

Carnarvon's death came just a couple of weeks after a public warning by novelist Marie Corelli that there would be dire consequences for anyone who entered the sealed tomb. Arthur Conan Doyle, the creator of Sherlock Holmes and a believer in the occult, announced that Carnarvon's death could have been the result of a "Pharaoh's curse."

One newspaper even printed a curse supposed to have been written in hieroglyphs at the entrance of the tomb, the translation being:

> *They who enter this sacred tomb shall swift be visited by wings of death.*

A complete fiction as it turned out, though one inscription found within the tomb did say:

> *It is I who hinder the sand from choking the secret chamber. I am for the protection of the deceased.*

However, an imaginative reporter added:

> *. . . and I will kill all those who cross this threshold into the sacred precincts of the Royal King who lives forever.*

Journalists determined to fuel the story of the Mummy's Curse reported other deaths attributed to the desecration of the pharaoh's tomb.

Five months after the death of Lord Carnarvon, his younger brother died suddenly. Another "casualty" was the pet canary of the tomb's discoverer, Howard Carter. The bird was apparently swallowed by a cobra on the day the tomb was opened. It was pointed out that the cobra was a traditional symbol of the pharaoh's power.

According to one list, of the twenty-six individuals present at the official opening of the tomb, six had died within a decade. However, many of the key individuals associated with the discovery and work on the tomb lived to a ripe old age.

As discoverer of the tomb, Howard Carter might have been considered a prime target for the curse. He had spent nearly a decade working inside it. But Carter didn't die until March 1939, just short of his sixty-fifth birthday and nearly seventeen years after first entering the tomb.

Even when some of the treasures of Tutankhamun went on tour overseas in the 1970s, some people still believed the curse might be at work. In September 1979, security guard George LaBrash had a stroke while watching over the mask of Tutankhamun at a San Francisco museum. He sued the city authorities for disability pay, claiming that the stroke was a job-related injury caused by a curse placed on anyone associated with the desecration of the tomb. The judge dismissed the claim.

THE CURSE OF PAPA DOC

Was retired U.S. Marine Corps colonel Robert Debs Heinl the victim of a voodoo curse?

From 1958 to 1963 Heinl served on Haiti as chief of the U.S. naval mission, while his wife, Nancy, studied the voodoo religion. Later, back in the United States, they began writing *Written in Blood— The Story of the Haitian People,* a history of the island. The book was widely expected to be openly critical of the ruling dynasty of François "Papa Doc" Duvalier. Some time later, after the death of Papa Doc, the Heinls learned from a newspaper published by Haitian exiles that a curse had been placed on the book by Papa Doc's widow, Simone.

Initial amusement turned to concern when a series of mishaps began to befall the book. First, the manuscript was lost on the way to the publishers. The Heinls prepared another copy and sent it off for binding and stitching, but the machine promptly broke down. A *Washington Post* reporter sent to interview the authors was struck down with acute appendicitis. Then Colonel Heinl fell through a stage when he was delivering a speech, injuring a leg. While walking near his home he was attacked and severely bitten by a dog.

On May 5, 1979, the Heinls were on holiday on St. Barthelemy Island near Haiti when the colonel dropped dead from a heart attack. His widow Nancy is reported as saying, "There is a belief that the closer you get to Haiti the more powerful the magic becomes."

ON THE ROCKS

A very powerful curse seems to hang over the Hawaiian volcano Mauna Loa.

Visitors to the beautiful island are warned by locals that the removal of volcanic rocks is likely to anger the goddess of the volcano, Pele, who is said to appear to warn of imminent eruptions. But it seems that some people simply won't be told.

During the summer of 1977 airline vice president Ralph Loffert of Buffalo, New York, his wife, and four children visited the volcano. Ignoring advice, they decided to take home a number of rocks as souvenirs.

Shortly after they returned home, Mauna Loa erupted. Within a few months one of the Loffert boys, Todd, developed appendicitis, had knee surgery, and broke his wrist. Another son, Mark, sprained an ankle and broke his arm; another son, Dan, caught an eye infection and had to wear glasses; and the daughter, Rebecca, lost two front teeth in a fall. In July 1978 the Lofferts sent the stones to a friend in Hawaii who was asked to return them to the volcano. But the disasters continued. Mark hurt his knee, Rebecca broke three more teeth, Dan fractured a bone in his hand and Todd dislocated an elbow and fractured his wrist again. Mark then confessed that he still had three stones. They were returned and the run of bad luck ended.

Mrs. Allison Raymond of Ontario, Canada, and her family also took some stones away from the volcano. She told reporters, "My husband was killed in a head-on car crash, and my mother died of cancer. My younger son was rushed to hospital with a pancreas condition that's slowly getting worse. Then he broke his leg. My daughter's marriage nearly broke up and it was only when I posted the rocks back that our luck improved."

Nixon Morris, a hardwood dealer from El Paso, Texas, was another who, in 1989, ignored warnings and took a Mauna Loa stone home. He promptly fell off his roof, his house was struck by lightning, and his wife fell ill with a mysterious infection that left her knee swollen. Morris then broke a hip and thigh when he fought with a burglar in their house, and his granddaughter fell and broke her arm in two places.

Morris had broken the Mauna Loa rock in two and given half to a friend. He said, "He brought the rock back to me after he wrecked four cars in less than two years and he'd never before had a wreck in his life."

In March 1981 Morris sent the rocks back to Hawaii.

John Erickson, a naturalist at the Volcanoes National Park in Hawaii, said he receives up to forty packages of rock a day from tourists who have returned home and experienced strange sequences of bad luck.

FRIDAY AFTERNOON CAR

James Dean died in 1955 when his Porsche Spyder sports car ran off the road. The car was taken to a garage, where it fell on a mechanic, breaking his leg. The engine was sold to a doctor who put it in his racing car, crashed, and died. In the same race, a car using the drive shaft from Dean's car crashed, and that driver was killed, too. When the car's shell was put on display, the car showroom burned down. It was exhibited again in Sacramento and fell off its stand onto a spectator, breaking his hip. The car was transported to Oregon, where it broke its mountings and smashed a shop window. In 1959 it is said to have broken into nine pieces while sitting on steel supports.

BABY CHAIR

Nine women working at a supermarket in Kent, UK, became pregnant over a ten-month period. They had all worked at checkout number 13.

THE KIMONO THAT BURNED DOWN TOKYO

A kimono successively owned by three teenage girls, each of whom died before she had a chance to wear it, was believed to be so unlucky it was cremated by a Japanese priest in February 1657. But as the garment was being burned a violent wind sprang up, fanning the flames and spreading them beyond control. The ensuing fire destroyed three-quarters of Tokyo, leveling 300 temples, 500 palaces, 9,000 shops, and 61 bridges, and killing 100,000 people.

MARTHA THE BOLT

Being married to Martha Martika had its positive and negative sides. The woman's first husband, Randolph, was struck down by lightning during a storm. Martha was devastated, but married again—to a young man called Charles Martaux. He, too, was killed by a lightning bolt. Martha fell into deep depression and sought help from a doctor. They fell in love and married. But he completed Martha's hat trick of electrifying bereavements when he stepped out into a thunderstorm and was struck and killed by lightning.

LIGHTNING STRIKES AGAIN

British cavalry officer Major Summerford was fighting in the fields of Flanders in the last year of the First World War when he was knocked off his horse by a flash of lightning. He was paralyzed from the waist down. He moved to Vancouver in Canada where, six years later, while fishing in a river, lightning struck him again, paralyzing his right side.

Two years later, he was sufficiently recovered to take walks in a local park. One summer's day in 1930, lightning sought him out again, this time permanently paralyzing him. He died two years later.

Zeus still hadn't had enough of Major Summerford. Four years later, lightning destroyed his tomb.

LIGHTNING LIKES ME

Kenny MacDonald could be forgiven for believing that lightning has developed a taste for him. During his thirty-four years working as a telephone line repairer, he was struck by lightning no fewer than three times.

"It wasn't really so surprising," he admits. "I was working in some pretty remote parts on telephone lines that were sometimes as much as thirty to forty miles long. Lines tended to come down in bad weather, and I would be sent out to repair the damage. People in remote communities rely on their telephones so perhaps I took some risks I shouldn't have.

"If you're up a pole at the end of a thirty-mile piece of wire in the middle of a thunderstorm, there is a real chance that lightning is going to find you. It was something of an occupational hazard. I was struck three, maybe four times in all those years, but the electricity passed straight through me and I have lived to tell the tale. If I'd been on the ground I would have been killed, but all I felt was a tickle and my hair stood on end. It leaves a coppery taste in the mouth."

Kenny thought that when he retired a few years back, his brushes with the brutal electrical force of nature were over. He was wrong.

On one wild and windy day he got up early with his son and set off in his car to go fishing. As they got near to where they planned to fish, the weather deteriorated. The clouds grew thicker, hail began to fall, and a thunderstorm erupted.

"Without warning a huge bolt of lightning struck us," says Kenny, who knows lightning when he sees it. "It blew a hole in the roof of the car. There was a huge bang, blue flames, and the smell and taste of copper. Our ears were ringing for twenty minutes.

"My son turned to me and said, 'Dad, was that what I thought it was?' I said yes and he said, 'Wow.'"

Kenny says they were saved from injury because the car's tires prevented the lightning from finding earth. If they'd had a flat tire they would almost certainly have been killed.

But Kenny wasn't about to let a little thing like a lightning strike get in the way of him and his beloved fishing. They limped on in their damaged car and settled down for a day's angling. Toward the end of the day Kenny managed to land a six-pound salmon. He caught it with a fly called "Thunder and Lightning."

Looking back on the events Kenny says he can't help but feel that lightning is following him around. He considers himself unlucky to have been struck so many times, but lucky to have walked away unscathed.

As for the future, he promises to be more careful around thunderstorms, particularly if he's fishing. "These new carbon fishing rods make fantastic lightning conductors," he says. "You can't be too careful."

And he has a word of comfort for those people who are frightened by the sound of thunder. "You never hear the lightning bolt that kills you," he says.

THE SCOURGE OF *LEO AND ME*

In the late 1970s, actor Michael J. Fox starred in a Canadian sitcom named *Leo and Me*. In the years that followed, Fox and three other colleagues from the series were diagnosed with the neurological disorder, early-onset Parkinson's disease. Is the fact that four people associated with the same show contracted the same relatively rare disease more than just coincidence? If it is then it may be possible to find a cure.

Early-onset Parkinson's disease has no universally recognized cause. "This is like a detective story," says Dr. Donald Calne of Vancouver's University of British Columbia Hospital, who treats Fox's three former coworkers. "We have to find the culprit. We have to consider a virus or a toxin."

Although he can't think of any environmental links, Don S. Williams, a former director on *Leo and Me* who developed Parkinson's in 1993, says the study of his case gives him hope. "I try not to get too excited," he says, "but I want to see if this leads to a cure."

8

HISTORY REPEATING ITSELF

The case of the moped rider killed in a collision with the same taxi that had already knocked down his brother on the same moped exactly a year before shows us that history has a nasty habit of repeating itself.

Why does it seem to be the unpleasant things in life that tend to come around a second time?

MURDER MOST SIMILAR

Two girls of the same age were murdered on the same day of the year, in the same place, 157 years apart. Mary Ashford, twenty, was found dead on May 27, 1817. The strangled body of Barbara Forrest, also twenty was found at the same location on May 27, 1974.

The bodies were found at points four hundred yards apart. Both girls had visited a friend earlier in the evening and they had both changed in order to go on to a dance. Both girls were raped before they were murdered. The deaths had occurred at approximately the same time of day, and in both cases attempts appeared to have been made to hide the bodies.

The men arrested for the murders were both named Thornton— Abraham Thornton and Michael Thornton. Both men denied the charge and both men were acquitted.

CROSSING OVER

A man whose daughter was killed by a train at a crossing four years before died at the same spot and at the same time. The train was being driven by the same man, Domenico Serafino.

Vittorio Vernoni, fifty-seven, used to drive back and forth to work several times a day over the Via Cartoccio crossing near Reggio Emilia, northern Italy, where his daughter Cristina, nineteen, was killed in 1991. The crossing, unmanned and without a protective bar—the legal norm for Italy's local railways—was equipped with two flashing lights and a bell to warn motorists of approaching trains. The crossing itself was situated near a bend. His daughter had been hit on the crossing on a bright winter morning and Vittorio drove over the crossing for the last time in his Renault on a sunny morning in November 1995. When the train driver, Domenico Serafino, spotted the car on the line and braked, it was already too late. His engine ploughed into the car and dragged it for several dozen yards.

Suggestions that Vittorio had decided to take his life at the place where his daughter had died were repudiated by his family and the train driver. Investigators said his death was accidental. Vittorio's son, Andrea, twenty-two, described it as an incredible, absurd fatal coincidence.

ONLY A FOOL MAKES THE
SAME MISTAKES TWICE

Though this chapter may so far have suggested otherwise, history doesn't always repeat itself merely to spite us. Sometimes when it drags things round again it turns out to be helpful. It certainly helped one of this book's authors, Martin Plimmer, when he was applying for his first job as a reporter on a local paper.

Martin hadn't learned much in life at that point, but somewhere along the line he had become aware of the maxim that we should learn from our mistakes. Even if it hadn't yet sunk in, here was a heaven-sent demonstration of the principle.

Martin was fresh out of journalism college, schooled in law, newspaper practice, libel, shorthand, and writing and interview techniques. His knowledge of religion, however, was sadly lacking. He learned how vital this could be at his first job interview.

It was an old-fashioned local newspaper in a tall and gloomily daunting Victorian building. The editor sat in a den of an office, seemingly sealed off from the hum and crackle of life. At the end of the interview, which was very formal, the editor said, "Now, I'd like to test your general knowledge." He stood up, walked across to his bookcase and picked out a thick blue book. He opened the book at random, looked the page up, and down, closed it, opened it again at random and said, "Ah yes, 'religion.' 1. What was the name of the mountain that Noah's ark came to rest on after the Great Flood?"

"Mount Sinai," Martin said. The editor frowned and asked a further nine questions. Martin got eight of them wrong. He didn't get the job.

When he got home that evening his father wanted to know how the interview had gone. He told him about the religious questions. "What questions?" asked his father.

"What was the name of the mountain that Noah's ark came to rest on after the Great Flood?"

"Ararat, of course. What did you say?"

"Sinai."

"No, that was the mountain where Moses was given the Ten Commandments." And so on through the ten questions.

A week later, Martin went for his second interview, at a different paper. It was a modern building this time, and once again the interview took place in the editor's office. It seemed to be going well but Martin's heart fell when the editor stood up, saying, "Now I'd like to ask you some general knowledge questions" walked over to his bookcase and took down a thick blue book. He opened the book at random and said, "Ah, this will do: religion. . . ."

Martin had to struggle not to answer the ten questions too fast. He deliberately got one wrong so as not to seem cocky. He got the job.

TRAIL BLAZER

Alleged arsonist Douglas Hunziker was the victim of an extraordinary series of coincidences, according to his attorney.

Witnesses at his arson trial said the former California firefighter trainee was present at nine of the ten blazes he is charged with setting, often helping to put out the flames or to evacuate people from the area. But the fact that Hunziker, twenty-two, was at nine fires in less than seven months didn't mean he started them, said his defense attorney Darryl Genis.

"It is entirely plausible," Genis told the jury, "that Douglas Hunziker is nothing more than a victim of extremely bad luck."

The series of fires began during the previous year's Fourth of July festivities. Hunziker called firefighters in the early morning hours to report a blaze in a trash container in front of his home. During the following few weeks Hunziker was closely associated with another nine fires. These included a blaze at a four-car garage at the home of his ex-girlfriend, shortly after the couple had broken up. Another fire broke out at the Comedy and Magic Club in Hermosa Beach, where Hunziker was working, and two days later the club was struck by arson that forced the evacuation of more than two hundred people.

Apparently Hunziker took an active part in the evacuation. Shortly afterward a blaze broke out at a local driving school that the defendant attended.

Hunziker was eventually arrested and charged with starting all ten fires. Remarkable coincidences do happen, but to find the defendant innocent of arson in this case, the jury would have had to believe in one of the most unlikely.

ERIC WILLIAMS STRIKES AGAIN

On July 6, 2003, Eric Williams blacked out at the wheel of his car, which left the road, ploughing through the front garden of sixty-two-year-old Gordon White and into his front living room.

The car was recovered, it took nine weeks to repair the house and almost a year later Mr. White added the final touches to the decorating.

Then on July 6, 2004, the same driver had another blackout on the same stretch of road and ploughed once again into Mr. White's living room. The same recovery man arrived to pull the car out for a second time.

"I'm going to make sure I'm not watching the TV this time next year," said Mr. White. "They say bad luck comes in threes."

ECHOES

Coincidence experiences can sneak up like ghosts, insinuating themselves into the thought process, and surreptitiously shifting you to a different place altogether. These strange and eerie reverberations, or echoes, can bring sadness, joy, and often bewilderment. Or they can bring understanding. Such an experience happened to Stephen Osborne, editor of the Canadian cultural and literary magazine *Geist*.

Osborne was relaxing in a bar with friends when the conversation turned to a man named Richard Simmins, a curator, art critic, and writer who had been an important force in Osborne's life twenty-five years before. Osborne regarded Simmins as his mentor, but hadn't seen him for seventeen years.

After a while the friends realized that the music in the bar had become very somber, and when they mentioned it to the bartender he shrugged and made a crack about it being a funeral parlor.

Later, in a reflective mood, Osborne arrived home and picked up a magazine to read from a pile in his bathroom. It fell open to a poem written in memory of Richard Simmins. "I understood at that moment that he was no longer alive," said Osborne. "The magazine was six months old; the poem, written by his daughter, would be how much older than that? I lit a candle to honor the man whom I had loved but had not seen since 1986."

LYRIC SUBSTANCE DISPUTED

It was conventional wisdom among both the rebellious young of the swinging sixties and their shocked elders, that John Lennon wrote the psychedelic song, "Lucy in the Sky with Diamonds," as a knowing reference to the drug LSD.

Lennon has always denied this, claiming the LSD association was a fluke coincidence. He said he had taken the title verbatim from a line written on a fantastic drawing by his four-year-old son Julian of his schoolmate Lucy O'Donnell. It had not occurred to him that the initials were the same as those of a mind-expanding drug much in the news at the time, until someone pointed it out to him after the release of *Sgt. Pepper's Lonely Hearts Club Band*, the record containing the song.

" 'Lucy in the Sky with Diamonds' . . . I swear to God, or swear to Mao, or to anybody you like, I had no idea spelled LSD," he told *Rolling Stone* magazine.

Needless to say his explanation was widely disbelieved. In fact the song caused something of a scandal at the time. Old people thought he was trying to pull the wool over their eyes and young people thought he was being frivolous. It was well known in 1967 that the Beatles were experimenting with drugs and shortly after the release of *Sgt. Pepper* they admitted to taking LSD. Also, the song contained a lot of what appeared to be hallucinatory acid trip imagery. Something like that just couldn't be an accident.

But in their excellent urban myth-debunking Internet site *Snopes.com*, Barbara and David P. Mikkelson, make a convincing argument for believing that Lennon's coincidence explanation was honest.

"He did not merely claim that the title was a coincidental invention of his own but offered a specific, external explanation of its origins; he provided this explanation at the time the song was released; he maintained the same explanation for the rest of his life; and his explanation is corroborated by others." Also the child picture stimulus is typical of Lennon's song writing muse. Other *Sgt. Pepper* songs by him were inspired by a Victorian carnival poster, a TV cornflake commercial, and a newspaper clipping about holes in Blackburn, Lancashire.

This was a man who had always been candid, sometimes too candid. Besides, what did he have to hide? He didn't deny his involvement with drugs. Why should he lie about the title of a song? Coincidence strikes again.

JIGSAW PUZZLE MEMORIAL

Stuart Spencer had been a widower for three years in January 2000 when his daughter gave him a present of a one thousand-piece jigsaw. She had found one of a paddle steamer on a lake, where Stuart and his late wife, Anne, had enjoyed many holidays. As he placed a piece to complete a figure in a wheelchair at the boat's stern, he saw it was his wife.

PIECING TOGETHER THE PAST

A similar puzzling coincidence was experienced by Jean Jones.

"My friend, who I have known since our school days, saw a jigsaw in a thrift shop and decided to buy it—it was a picture of a lovely garden.

"She took it home and started to do the puzzle. As she fitted the last piece, she recognized me and my late husband, Cyril walking past the garden.

"She didn't tell me about this amazing coincidence until she had stuck it all together on a board for me to hang as a picture. At first I couldn't believe it, and even now, after my husband has been dead more than five years, I still think this is a chance in a million. Since then my friend and I have seen many puzzles with the same picture but without my husband and me in it."

DOUBLE EXPOSURE

Just before the outbreak of the First World War, a German mother took a photograph of her small son. She left the single-plate film to be developed in Strasbourg in France, but the start of the war made it impossible for her to collect it. Two years later, in Frankfurt, she bought another film to photograph her newborn daughter. When the film was developed it turned out to be a double exposure. The picture underneath was that of her small son, taken in 1914.

A CARPENTER'S REQUIEM

A guitar being played by Andrés Segovia split apart during a Berlin performance at the same time as his friend, the craftsman who had made the instrument, died in Madrid.

SCHOOL SIGN

Objects and associations sometimes come together with such apparent perspicacity you feel sure they must be trying to say something to you. Just what they're trying to say, though, is not so clear.

This happened to Martin Plimmer when he and his eleven-year-old son were visiting secondary schools that the boy might attend. They were looking around the geography department of one particular school when Martin picked up one of the textbooks and said, "Geography was one of the few subjects I liked at my school. It's straightforward and covers a lot of interesting ground." Opening the book at random he realized he was looking at a page showing a large-scale map of a town with the secondary school he had attended there nearly forty years before clearly marked.

The effect of this sudden correlation of idea and image was to produce a pang of emotion and some confusion in Martin. His own education all those years ago, his son's about to start, and all their associated hopes and worries, seemed to be suddenly condensed into this one symbolic moment. What was this, if not a sign? And if a sign, what did it mean?

Could it signify that the school was ideally suited to his son, or conversely, as he had such a miserable time at the school, that it was the wrong school for his son? Opposite notions they might be, but both ideas were oddly compelling. That was until Martin recovered his senses a few seconds later by concluding that this wasn't an omen at all. There could be no answer to the question of the coincidence's significance. It was merely a demonstration of how even the most skeptical of us can be rendered momentarily gullible when the paranormal appears to nudge us in the ribs.

"YELLOW BIRD" DUET

There was a big round control knob on the side of Tom's parents' record player/radio (the knob toggled between the two) that just begged to be turned. The year was 1961 and eleven-year-old Tom was at home in Illinois listening to one of his first vinyl records, the Arthur Lyman Group's instrumental version of "Yellow Bird," which at the time was riding high in the charts.

Tom was slightly worried that the player might be damaged if the control was turned in the middle of a track, but he risked a turn anyway. To his alarm there was no change in the sound. He thought he'd broken the player.

In fact the song was being broadcast simultaneously by a radio station his parents' set happened to be tuned into. Tom switched back and forth in panic for a while and eventually detected a slight time difference appearing between the two. This was due to minute differences of speed between the two turntables—his parents' and the one at the radio station.

TORNADO

A tornado that wrought havoc in East St. Louis, Illinois, on May 27, 1896, seems to have been hell-bent on destroying the memory of the great engineer James B. Eades.

Eades had built the bridge over the Mississippi River that bore his name. A memorial window in the local Mount Calvary Episcopal Church has the inscription: "In memory of James B. Eades. Born May 23, 1820. Died March 8, 1887."

The 1896 tornado tore down the east section of the Eades Bridge at precisely the same moment it blew in the memorial window. It was the only window in the church to be damaged.

SIMULATED COINCIDENCE

A U.S. intelligence agency was planning an exercise simulating an aircraft crashing into a tower building, on the very morning of the September 11 terrorist attack on the Twin Towers of the World Trade Center.

Officials at the National Reconnaissance Office in Chantilly, Virginia, had scheduled an exercise in which a small corporate jet would "crash" into one of the four towers at the agency's headquarters building. Terrorism was not a factor in the scenario. The aircraft was to have developed mechanical failure.

The NRO building is about four miles from the runways of Washington Dulles International Airport. Agency chiefs came up with the scenario to test employees' ability to respond to a disaster. No actual plane was to be involved. To simulate the damage from the crash, some stairwells and exits were to be closed off, forcing employees to find other ways to evacuate the building. "It was just an incredible coincidence that this happened to involve an aircraft crashing into our facility," said spokesman Art Haubold. "As soon as the real-world events began, we canceled the exercise."

American Airlines Flight 77—the Boeing 767 that was hijacked and crashed into the Pentagon—took off from Dulles at 8:10 a.m. on September 11, less than an hour before the exercise was to begin. It struck the Pentagon around 9:40 a.m., killing 64 aboard the plane and 125 on the ground.

In another extraordinary twist, the pilot of Flight 77, Charles Burlingame, a former navy flyer, had, as his last navy mission, helped prepare Pentagon response plans in the event of a commercial airliner hitting the building.

MOON RIDDLE

Obscure evidence leads to imaginative interpretation, often wildly contradictory to the facts. A good example of this process is a story from the early history of science. In 1610 Galileo thought he had detected two moons orbiting Saturn (they were actually the planet's rings).

He conveyed this information by letter to his contemporary, the German astronomer and mathematician Johannes Kepler, but in order to keep the information as secret as possible he couched the crucial sentence as an anagram: smaismrmilmepoetaleumibunenugt-tauiras. Taking two of the Medieval "U"s to be Latin "V"s, the unscrambled message reads *Altissimum planetam tergeminum observavi*, which translates as "I have observed the highest of the planets—Saturn—three-formed."

Incredibly, Kepler managed to interpret the anagram another way, which also made perfect sense. Kepler read Galileo's inscrutable message as *Salve umbisteneum geminatum Martia proles* ("Hail, twin companionship, children of Mars"), a prediction that Mars has two moons. This was a conclusion Kepler had come to himself so Galileo's message would have endorsed his own calculations.

Despite the most rigorous standards our minds see everything through the filter that preoccupies them most. It is interesting to note that Kepler's interpretation wasn't quite a perfect coincidence; his phrase requires an extra letter. He must have thought that too small a flaw to spoil a desired result.

BAD LUCK AT SEA

Those who find even a short ferry trip an unsettling experience will have the utmost sympathy for the crew and passengers involved in this extraordinary tale of maritime misfortune.

On October 16, 1829, the schooner *Mermaid* left Sydney bound for Collier Bay on the northwest coast of Western Australia. On board were eighteen crew and three passengers. The captain's name was Samuel Nolbrow.

Four days later a gale in the Torres Straits blew the vessel onto reefs, marooning all on board for three days until they were rescued by the barque *Swiftsure*.

Five days later, the *Swiftsure* was caught in a strong current off New Guinea and was wrecked on rocks, though the combined crews and passengers made it safely to land.

Eight hours later they were all rescued by the schooner *Governor Ready*, but within three hours, the timber cargo caught fire sending everyone rushing for the lifeboats.

The cutter *Comet* picked them all up without loss of life. A week later, the *Comet* ran into a squall and snapped her mast. Her crew took the only serviceable boat, leaving the previously rescued crews and passengers to fend for themselves. For eighteen hours they clung to the wrecked ship until they were eventually rescued by the packet *Jupiter*.

By a further strange coincidence, a passenger on the *Jupiter* was an elderly woman from England who was on her way to Australia to look for her son who'd been missing for fifteen years. She found him—he was a crewman from the *Mermaid*.

PLUM PUDDING

A certain M. Deschamps, when a boy in Orleans, was once given a piece of plum pudding by a M. de Fortgibu. Ten years later he tried to buy a slice of plum pudding in a Paris restaurant but was told that it was already sold—to M. de Fortgibu.

Many years afterward, M. Deschamps was enjoying some plum pudding with friends and remarked that the only thing missing was M. de Fortgibu. At that moment the door opened and an old, somewhat confused man walked in. He apologized, explaining that he had got the wrong address. It was M. de Fortgibu.

DEATH ON THE BRIDGE

In February 1957, Richard Besinger, aged ninety, was run over and killed while walking in the middle of the Big Lagoon Bridge in Eureka, California. Two years later his son Hiram was killed on the same bridge when a timber truck overturned on him. Six years after that, his great-grandson David Whisler, aged fourteen, was killed by a car—on the same bridge.

GOOD AND BAD INTELLIGENCE

The bulky National Commission report into the circumstances surrounding the 9/11 terrorist attack has the benefit of hindsight, but what shines out amid all the confusion, unpreparedness, inefficiencies, and wrong responses of intelligence and law enforcement agencies in the events leading up to the attack, are those few instances when officials judged the situation exactly right. Kenneth Williams, an FBI agent in Phoenix, Arizona, sent a memo to the agency head-quarters on July 10, 2001. He had noted the arrival of Middle East-ern students at a local flight school. His suspicion: terrorists might be trying to infiltrate the civil aviation system. Williams's memo was ignored.

On August, 16, FBI agents arrested Zacharias Moussaoui for suspicious activity at a flight school. Moussaoui had shown little in-terest in learning to take off or land. The arresting agent actually wrote that he was the "type of person who could fly something into the World Trade Center." Another FBI agent on the case speculated that a large aircraft could be used as a weapon. Security officials gave briefings at the White House in the summer of 2001 in which Pres-ident Bush was informed that bin Laden's terrorist network might try to hijack American planes, that attacks might take place in the com-ing weeks, that they would be spectacular and designed to inflict mass casualties, and that the FBI was aware of terrorist preparations, including the surveillance of buildings in New York.

Such prescience is the product of good analysis of evidence and sound intuition, though there may be a measure of coincidental guesswork in there, too. Whatever, had the government taken the warnings of these agents seriously, the 9/11 terrorist attack might have been averted.

. 10 .

NAMES

Just as we take delight in learning that an acquaintance shares our birthday, we can gain as much pleasure from learning that he or she shares our name. But if someone shares our name and our birthday—we might have a problem. As was the case in the story of the two Belindas.

The women were born on January 7, 1969, and both were named Belinda Lee Perry. They became aware of each other when both joined the same library—and Belinda One's card was canceled, causing problems for Belinda Two. Later they were both investigated for fraud when both made applications for student grants.

When Belinda Two moved home, Belinda One got struck off the electoral roll. After finishing school, both went to work as clerks in public service for about eighteen months, then both worked at the same university for about the same length of time. They finally met when they both enrolled as students at the university. They found they had a lot in common.

BIRDS OF A FEATHER CRASH TOGETHER

Margaret Bird's car was involved in a collision with another car and a van. All three drivers were called Bird.

PECKING ORDER

On the list of employees of a duck farmer were two people called Crow, four called Robbins, a Sparrow, a Gosling, and a Dickie Bird.

CHARMED NAME

Life jackets, lifeboats, and emergency rations are all useful things in the event of a shipwreck, but having the right name can also help. Hugh Williams was the only survivor of a vessel that sank in the Straits of Dover on December 5, 1660. One hundred and twenty-one years later to the day, another shipping disaster in the same waters claimed the lives of all on board, except a man with the seemingly charmed name of Hugh Williams. On August 5, 1820, when a picnic boat capsized on the Thames, all drowned with the exception of a five-year-old boy—Hugh Williams. On July 10, 1940, a British trawler was destroyed by a German mine. Only two men survived, an uncle and nephew, both named Hugh Williams.

THESE YOUR DWARFS, MADAM?

A series of shotgun raids in Barcelona, Spain, turned out to be the work of a gang of seven dwarfs masterminded by a tall, good-looking blond whose name was Nieves, which means snow in Spanish.

LIKE DRAWN TO LIKE

A golfer watched his perfect drive collide mid-flight with another ball—a recovery shot by another player in the opposite direction. Astounded by the coincidence, the golfer and the other player ran to the collision point to introduce themselves. They were both named Kevin O'Brien.

GREENBERRY HILL

A magistrate was murdered in 1678. Three men were arrested, tried, and found guilty of the killing of Sir Edmund Godfrey. They were hanged at Greenberry Hill. Their names were Green, Berry, and Hill.

LETTER FOR MR. BRYSON

George D. Bryson was on a business trip to Louisville in Kentucky and had booked to stay at the local Brown Hotel. He checked in, received the keys to room 307, and asked if any letters had arrived for him. He was handed a letter addressed to Mr. George D. Bryson, room 307. But the letter wasn't for him. It was intended for the previous occupant of room 307—another George D. Bryson.

W.C.

During World War II, Winston Churchill visited a naval base to see a demonstration of the "Asdic" antisubmarine defence system.

He was taken to an area of submerged wrecks and watched as a target was located on the seabed and a depth charge dropped. There was an enormous explosion and the ship swayed sickeningly. A few moments later a door floated to the surface bearing the letters "W.C."

"The navy always knew," said Churchill, "how to pay proper compliments."

DOUBLE CHANCE

It had to be a chance in a million. Motorcyclist Frederick Chance collided with a car being driven by none other than Frederick Chance. Neither Fred nor Fred was, by chance, seriously hurt.

GUESS WHO I RAN INTO

Dr. Alan McGlashan recalls the coincidence that happened to his stepson Bunny as he was driving home to his seaside cottage at about 2 a.m. one morning.

"A man driving out of a side road ran straight into his car. It was on the outskirts of a town and the town sheriff happened to be nearby on his night rounds. He took out his notebook and said to the driver of the other car, 'Your name please?' The man said, 'Ian Purvis.' My stepson, who likes to play things deadpan, said nothing. The policeman then turned to Bunny and said, 'And your name, sir?' And Bunny said, 'Ian Purvis.' 'Look here,' said the bobby, 'this is no time for silly jokes.' But it was the truth. One Ian Purvis had run into another Ian Purvis."

TWIN DEATHS

Seventy-one-year-old Finnish twin brothers were killed in identical bicycle accidents along the same stretch of road just two hours apart. "Although the road is a busy one, accidents don't occur every day," said police officer Marja-Leena Huhtala. "It made my hair stand on end when I heard the two were brothers, and identical twins at that. It's an incredible coincidence. It makes you think that perhaps someone upstairs had a say in this."

PUZZLE PATTERN

Keen mah jong player Jill Newton advertised for a mah jong set in a newspaper and got a response from a family in Gillingham. On the way to look at the game, her car was involved in a minor traffic accident. She exchanged addresses with the other driver. The name of his house was Mah Jong.

A TALE OF TWO TURPINS

An air mail letter addressed to Miss S. Turpin was delivered in 1955 to the home of the Marquesa de Cabriñana, at Calle Goya 8, Madrid, and opened by Sallie Turpin, the English governess to the marquesa's children. But Sallie couldn't make head nor tail of the letter, which referred to people she had never heard of and was signed "Your loving Mom." Enquiries with the caretakers of neighboring houses revealed that an American girl named Susie Turpin lived at Calle Goya 12.

The wrongly addressed letter created a friendship between the two girls, who even went on vacation together. Sallie (now Sallie Colak-Antic) says, "I have often wondered what the odds were of two girls, one from England and the other from America, with the

same surname and first initial, finding that they were living in a foreign country two doors away from each other."

TOILET TALK

For too long controversy has raged among scholars and drunkards
about the origin of the word "crap." It's time to clear it up.

The popular notion is that the word derives from the name of
Thomas Crapper, who ran a highly successful sanitary ware business in England in the nineteenth century and is said to have invented the flush lavatory. It does not. The connection is pure
coincidence.

The word probably derives from the Middle English "crappe," or
"chaff" (possibly related to the Dutch "crappen," to break off).
From there it was applied to other unwanted residues.

It had long settled into its vulgar use by the time Thomas Crapper was born in 1836, and had made its first appearance in a slang
dictionary in 1846, when he was but ten years old, far too young even
for an able person such as him to have founded a plumbing legend.

It's easy just to say "coincidence" and wrap it up in a one-word
explanation, but there's a price to pay in doing so, for we rob the
story of its romance. Who's to say, given its associations, that his
name played no part in the boy's choice of career? Who's to say that
the young Crapper was not predisposed to see his destiny written, if
not in the stars, then in something more reliably solid upon which a
toilet empire might be built? Certainly some sense of mission must
have sped the boy south from his native Yorkshire at the age of fourteen, to be apprenticed to a master plumber in Chelsea, London.

Thomas Crapper joined a celebrated band of professionals
whose names bolt them to their jobs, including Mr. Rose, the Birm-
ingham School gardener; Dr. Zoltan Ovary, the New York gynaecol-
ogist; A. Moron, the Virgin Islands Commissioner for Education; the
Reverend God of Congaree; Doctor Doctor the doctor; Mrs. Screech
the Canadian singing teacher; I. C. Shivers the iceman; Lawless and

Lynch the lawyers from Jamaica, Queens; Shine Soon Sun the Texan geophysicist; Wyre and Tapping the New York private detectives; Mr. Vroom the South African motorcycle dealer; Ronald Supena, the Philadelphia lawyer; Preserved Fish Jr., the Massachusetts whale oil dealer; Major Minor of the U.S. army; Justin Tune, the Princeton chorister; Groaner Digger, the Houston undertaker; and Mr. Vice of New Orleans, who, according to the *International Herald Tribune*, was arrested 890 times and convicted 421.

A lesser man might have changed his name to Thomas Powder-My-Nose, but Crapper was never shy. He proclaimed the name on the roof of the company he founded. In fact he was something of a self-publicist and pioneered the bathroom showroom, in which his wares were displayed to the street, causing refined ladies to faint at the vulgarity of it all. None of this prevented a flush of royal commissions from two kings in succession. He was the Richard Branson of his day.

He may, however, have inspired the American use of the word *crapper* for toilet. It's said that American soldiers stationed in Britain during the First World War adopted the expression after seeing the words "T. Crapper—Chelsea" printed on British cisterns.

The Existential Press Officer

Journalists covering the story of post-Afghanistan prisoners holed up in the U.S. detention center at Guantanamo Bay, Cuba, often note how Kafkaesque their plight is. This is a reference to the early twentieth-century Czech author Franz Kafka, who was concerned with the alienation and dehumanization of modern man. Kafka's famous novel *The Trial* is about the incarceration of Josef K, a man who has done no wrong and whose alleged crime is never revealed to him, though he is subjected to endless interrogation.

As if to reinforce this impression, the U.S. Navy spokesman appointed to answer journalists questions about Guantanamo Bay was Lieutenant Mike Kafka.

DOUBLE DIX

Theater people, like writers, are always on the lookout for good stories and so tend to notice and remember coincidences. Actor Christopher Eccleston told the author Martin Plimmer a story about the American actress Sophie Dix, who lost her purse and later found another purse in the street belonging to another woman, also named Sophie Dix.

DOUBLE TAKE

Englishman David Webb's trip to America gave him a chance to discover himself.

On vacation in Florida he decided to hire a car and drive across the Everglades. Stopping for a break, he entered a roadside diner, ordered a coffee, and fell into conversation with the only other customer in the place,

As they talked, a series of extraordinary coincidences emerged. Both men were named David Webb. They were both born on the same day, December 24, of the same year. David Webb from England lived in Camden Passage in north London behind a row of antique shops. The David Webb in America had formerly lived in London where he owned one of those Camden Passage antique shops.

D-DAY CROSSWORD

On May 3, 1944, Allied war leaders became alarmed when a four-letter code word for the D-day beach assigned to the 4th U.S. Assault Division appeared as an answer in the crossword of British newspaper the *Daily Telegraph*. The word was "Utah."

Was this just a coincidence? In previous months the words "Juno," "Gold," and "Sword" (all code names for beaches assigned to the British) had appeared, but they were common words in crosswords. Then on May 22, 1944, came the clue "Red Indian on the Missouri (5)." The solution was "Omaha"—code name for the D-day beach to be taken by the 1st U.S. Assault Division. It was followed by solutions including "Overlord," the code for the whole D-day operation and "Mulberry," the code for the floating harbors used in the landings. Finally on June 1, the solution to 15 Down was "Neptune," the code word for the naval assault phase.

It was just four days before D-day. Two highly suspicious officers working for MI5, the British secret service, were dispatched to interrogate the *Telegraph*'s crossword compiler Leonard Dawe, who was also the headmaster of London's Strand School. "They turned me inside out," Dawe said later.

The MI5 men concluded that Dawe was not a secret agent and that the sequence of compromising words was an extraordinary fluke—a coincidence. The story, with this explanation, became just another piece of the historical folklore of D-day.

It was to be another forty years before the real explanation emerged. Property manger Ronald French, responding to an article retelling the story in a newspaper, revealed that it was he, as a fourteen-year-old pupil at the school, who had inserted the words in the puzzles.

It had been his crossword-compiling headmaster's habit to invite pupils into his study and ask them to suggest "answer" words to fit the blanks in the crosswords. He would then create clues for them. French said that during the weeks before D-day he had overheard these words used by Canadian and American soldiers, who were camped close by the school, awaiting the invasion. But he insisted that the names were common knowledge. It was just times and places that were secret.

PARALLEL LIVES

Imagine that there is not one universe, but countless. In each, history unravels in similar, but not identical fashion. Time passes at the same rate in each of these universes but is not synchronized—the differences are staggered.

As a rule these universes are self-contained and have no influence on each other. But sometimes the fabric that separates these parallel universes is damaged and events in one existence leak into another.

That seems a perfectly plausible explanation for at least some of these following stories of parallel lives. Surely it isn't just coincidence?

ONE WANDA TOO MANY

The odds against the coincidences that link the lives of the two Wanda Marie Johnsons are staggering. This story of almost duplicate lives was first reported in the *Washington Post* on April 20, 1978.

At the time, Wanda Marie Johnson of Adelphi, Maryland, in Prince Georges County, worked as a baggage clerk at Union Station in Washington.

The second Wanda Marie Johnson lived in Suitland, Maryland, also in Prince Georges County, and worked as a nurse at DC General Hospital in Washington.

Both women were born on June 15, 1953 and were former District of Columbia residents who moved to Prince Georges. Both had two children and both owned 1977 two-door Ford Granadas.

The eleven-digit serial numbers of their cars were the same except for the last three digits. Their Maryland driver's licences were identical because a computer determines each licence number by name and birthdate.

As a result, Wanda Marie Johnson of Adelphi became the victim of medical record mix-ups and harassment for payment of debts she didn't owe, received telephone calls from strangers, and was told by Maryland Department of Motor Vehicle officials that she must wear glasses while driving. It was the Wanda from Suitland who was shortsighted.

The problems had begun when both women still lived in the District of Columbia—one on Girard Street, the other on New Jersey Avenue.

They both had babies at Howard University Hospital and attended the same Howard clinic; Wanda of Girard Street realized something was wrong when doctors at the clinic began referring to the other Wanda's medical information.

All her efforts to trace her namesake were unsuccessful.

The two Wandas were eventually brought together by a newspaper reporter. They got on fine, but neither was prepared to change her name.

UMBERTO, DEUXBERTO

Umberto the restaurant owner bore a striking resemblance to King Umberto I of Italy. He had the same name and he, too, was born on

March 14, 1844, in the same town. He also married on 22 April, to a woman also called Margherita. His son, just like the king's, was called Vittorio, and on the day of the king's coronation in 1878, he opened his restaurant.

All this wouldn't have had a lot of significance had the king and the restaurant owner never met, but, just like in a fairy story, the two Umbertos' paths did cross. They were to find out that their lives had more similarities than just these simple facts.

The king had come to Monza, near Milan, to present prizes to athletes. The night before the tournament he and his aide went for dinner at Umberto's restaurant. As he sat there the king noticed the physical resemblance between Umberto and himself. He called the man over and as they swapped vital statistics the two men began to marvel. The list of similarities was teased out, savored and remarked upon and then, to top it all, they discovered they had both been decorated for bravery on the same day, on two occasions, in 1866 and in 1870. The king determined at that point to make the restaurateur a Cavaliere of the Crown of Italy and invited him to the athletics tournament the following day.

Like many a fairy tale, this one also has a violent ending. There was no sign of Umberto the next morning and when the king asked after him he was told the restaurateur had died that day in a shooting accident. Saddened, King Umberto declared that he would attend his funeral. It was never to be. The king was killed that day by three shots from the pistol of the assassin Gaetano Bresci.

NOT FEELING HIMSELF

We like to think of coincidences as benign or amusing, but often the opposite is true. Here are the ingredients of a nightmare: two patients at the same hospital, with the same Christian name and surname, both suffering from brain tumors and nobody realizing the link between them.

In this case a simple coincidence caused weeks of fear and misunderstanding for two families, though as luck would have it, a second coincidence made the resolution of the problem possible.

When Sheila Fennell's son Stephen began a worsening sequence of depressions and blackouts, and eventually suffered a fit, his doctor arranged for him to have an electroencephalogram, or EEG test. The family then had a long and tense wait for the results. Eight weeks went by, the results hadn't arrived, and Stephen's health was rapidly deteriorating. When he suffered another fit, Sheila called an ambulance and he was admitted to the hospital. But there was still no EEG result. Another ten days went by. The hospital kept telling Sheila he'd had a scan on August 31 that had shown he was okay. Stephen hadn't had a scan on the thirty-first.

Puzzled and increasingly desperate, the Fennells began clutching at straws. Sheila remembered that her husband had once mentioned that there was a young man at the power station where he worked who was also named Stephen Fennell. It seemed unlikely but he agreed to ask that Stephen if he'd ever had a brain scan. Sure enough he had, and his family were also desperate, but for a different reason: they had received results telling them that their Stephen had a very large and inoperable brain tumor.

Of course, the results had been sent out to the wrong patients, but this brought no relief to Sheila because her son now seemed to be fighting for his life. Fortunately the tumor turned out to be benign, and though large, it was possible to drain it. Stephen started to get better and after two years began to look more like his old self.

The mix-ups weren't over though. Three years later he got a letter asking him to go to the specialist urgently. The doctor asked him if he was still getting the headaches he'd reported in March. This time they were able to direct the doctor to examine his files carefully. Once again the two files had been muddled together.

Knowing what had happened before, this was something they could easily sort out, but Sheila often wonders how the story would have ended if her husband hadn't, by coincidence, worked alongside a man called Stephen Fennell.

MARTIN GUERRE

In the summer of 1557, Frenchman Martin Guerre returned to the village of Artigat after eight years away from home fighting in the wars that culminated in August that year, in the battle of St. Quentin, in which the English and Spanish defeated the French.

Guerre, a wealthy landowner, soon settled back into family life with his wife Bertrande and his relatives. Life was comfortable, with good rents coming in from his properties. To his delight his first child, a daughter, was born.

But some among his family, in particular Guerre's younger brother Pierre, thought that the war had changed him a little more than was credible. He seemed to have forgotten expressions common to his dialect, he showed no interest in his once engrossing interests, swordplay and acrobatics, and worse, he sold off some parcels of land that had been in the family for generations.

Others in the village believed he was genuine and Bertrande, who realized he was an impostor, but preferred this warm, charming man to her rather heartless (and what's more absent) husband, insisted it was so.

A few years later Pierre, who had never stopped campaigning against the man, succeeded in having him arrested as an impostor. A long and complicated court case ensued with many witnesses called. The case went to appeal in Toulouse and due to the impossibly contradictory nature of all the many testimonies, the court was on the verge of giving the defendant the benefit of the doubt, when a one-legged man on crutches hobbled into the courtroom. Bertrande collapsed. It was the real Martin Guerre, who (perhaps having got wind of the case) had returned from Spain, where he had spent the last few years living a new life.

The truth of the story was that during a battle in Flanders, Martin Guerre had lost a leg and been left for dead on the battlefield. Another soldier, Arnaud du Tilh, came across Guerre and, realizing they looked alike, down to having the same twisted fingernail, the same four warts on the right hand and an identical scar on the forehead, concluded that his material needs would be solved if he took the place of the man he assumed to be dead. Arnaud du Tilh, a clever rascal and skilled actor, walked into the village in 1557 and took over Guerre's home.

For du Tilh there remained only the gallows. Bertrande, spared the rope because of her gender, was forced to watch the execution of the man she had grown to love and return to live with the cold man who had deserted her to live in Spain.

TWIN SISTERS

Tamara Rabi and Adriana Scott couldn't understand why strangers kept approaching them in the street claiming to know them.

The girls, both students at nearby universities in New York, grew increasingly puzzled. It happened so regularly that they started to believe they must have doubles or doppelgängers in the neighborhood.

The explanation was that Tamara and Adriana were in fact identical twin sisters, born in Mexico in 1983 and given up for adoption at birth. They were handed over to different sets of adoptive parents who had flown down from America.

By an extraordinary coincidence they ended up living just twenty-five miles apart, but their upbringings were entirely different. Tamara was adopted by a Jewish couple who lived close to Central Park in Manhattan; Adriana grew up in a Roman Catholic family, in the Long Island suburb of Valley Stream.

They were eventually brought together on their twentieth birthdays. Justin Latorre, a friend of Tamara's, happened to be invited to Adriana's party. She was startled by the resemblance between the two girls and decided it could not be mere coincidence. She insisted that the pair make contact.

Both girls knew that they had been born in Mexico and adopted, but knew nothing of each other's existence. Adriana sent Tamara a photograph of herself.

"I had seen her face every time I looked in a mirror," said Tamara. "I just felt that I had known her all my life." Both girls put some pointed questions to their adoptive parents, and the truth emerged.

A few days later, the young women met for the first time and spent the following weeks getting to know each other. They learned that both had lost their adoptive fathers to cancer, Adriana when she was eleven, and Tamara only the previous autumn. Both had crashed into plate glass doors in childhood and both were musical and loved to dance. Both even had photographs of themselves as infants wearing identical Minnie Mouse romper suits.

"Then we discovered that we both had the same recurring dream," said Tamara. "It involves a very loud noise, then everything going quiet, and then very noisy again, and it was always very scary," said Adriana. "It must have been something that happened to us when were together in our mother's womb," said Tamara.

DOUBLE ACT

Albert Rivers and Betty Cheetham shared a table with another couple for dinner at the Tourkhalf Hotel in Tunisia in early 1998. The other couple introduced themselves—Albert Cheetham and Betty Rivers. All were in their seventies. Other similarities emerged. Both couples were married on the same day and at the same time. Both had two sons, born in 1943 and 1945. Both had five grandchildren and four great-grandchildren. The Bettys had worked in post offices in their home towns while their husbands had been carriage bodybuilders in railway workshops. Neither woman could show her engagement ring as both had lost them, but they did have identical watch bracelets, which had repairs to the same links.

IF THE KEY FITS

Sales representative Robert Beame could certainly be forgiven for believing he had stepped into a parallel universe back in 1954 when he was traveling in Iowa.

Leaving his hotel one morning, he returned to his car and began his journey to his first business appointment of the day in a nearby town. A few miles out, he pulled over to look at some papers in his briefcase, which was still lying on the passenger seat where he had left it the previous evening.

Pulling out some documents, he found to his astonishment that they did not belong to him. Checking the contents of the briefcase more carefully he realized it was not his. Had someone swapped the briefcases? Nothing else seemed to be out of place in the car. It did not appear to have been broken into. He then checked the glove compartment and found other items that did not belong to him. Beame eventually drew the inevitable conclusion that he had somehow got into the wrong car.

He drove back to the spot where he had parked the previous night and standing there was a man in front of an identical car that proved to be his. The coincidences continued. The owner of the car he had taken in error was a man he had roomed with in college seven years earlier. They had not been in touch since. When the two men examined their cars they found they were identical down to the hood ornament that each had specially ordered.

"As I recall my key fitted his car but his key did not fit mine," says Robert.

LINCOLN AND KENNEDY

A study of the lives and violent deaths of Presidents Abraham Lincoln and John F. Kennedy reveals some remarkable coincidences. Eminent mathematician Professor Ian Stewart is not convinced that the parallels amount to something "beyond coincidence," but he concedes that not everything that happens in the universe can be understood or explained. The circumstances, he agrees, are certainly very compelling.

Many books, newspapers, and Internet sites have catalogued the Kennedy/Lincoln coincidences. Most, in their enthusiasm, contain inaccuracies, distortions and exaggerations. But most agree on the following details:

- Lincoln was elected president in 1860. Exactly one hundred years later, in 1960, Kennedy was elected president.
- Both men were involved in civil rights.
- Both were assassinated on a Friday, in the presence of their wives.
- Both men were killed by a bullet that entered the head from behind.
- Lincoln was killed in Ford's Theater. Kennedy met his death while riding in a Lincoln convertible made by the Ford Motor Company.
- Both men were succeeded by vice presidents named Johnson who were Southern Democrats and former senators.
- Andrew Johnson was born in 1808. Lyndon Johnson was born in 1908, exactly one hundred years later.
- Assassin John Wilkes Booth was born in 1839. Assassin Lee Harvey Oswald was born in 1939, one hundred years later.
- Both assassins were Southerners. Both were murdered before they could be brought to trial.
- Booth shot Lincoln in a theater and fled to a barn. Oswald shot Kennedy from a warehouse and fled to a theater.

But even this consensus view contains at least one often-repeated glaring error—Booth was, in fact, born in 1838, not 1839.

The most detailed investigation into the Kennedy/Lincoln coincidences has been conducted by Australian writer, Ken Anderson. In his book, *The Coincidence File*, he agrees that the coincidences involving the two presidents are often inaccurately reported, but says he believes he has uncovered parallels that more than compensate.

Among the points he makes are:

- Both Oswald and Booth shot their victims in the head. According to Anderson, assassins in public places tend to aim at the heart or other vulnerable body parts when using a gun. Examples include Charles Guiteau, whose bullet struck President James Garfield in the pancreas on July 2, 1881, and Arthur H. Bremner, who shot Alabama governor George Wallace five times in the body on May 15, 1972.
- Both Lincoln and Kennedy liked to be able to travel openly around the country and disliked being surrounded by guards. Kennedy had ordered the top to be removed from his Lincoln Continental so that he and his wife Jackie could be more easily seen during the cavalcade through Dallas. Lincoln had also been careless about his security. When Booth entered Ford's Theater, he found the presidential box unguarded. John Parker, the White House policeman charged with protecting the president, had left his side several times, once to get a drink and on another occasion to get a better view of the production.
- At the time of their assassinations, both presidents were accompanied by their wives and another couple. In each case the other man was wounded by the killer. John Connally, the governor of Texas, and his wife were traveling in the car with President Kennedy. A bullet passed through Connally's body and emerged to hit him in the wrist, before entering his thigh. Major Henry Rathbone and his fiancée Miss Clara Harris were in the box at the theater with President Lincoln. Rathbone attempted to tackle Booth after the shooting and was stabbed in the arm with a hunting knife.
- Both presidents were sitting beside their wives at the time of the shootings. Neither woman was injured. Both women cradled their dying husbands' heads in their hands. Each had to

wait while doctors made desperate but unsuccessful attempts
to save their husbands. Both women had married at the age
of twenty-four. Both had three children and both had a child
die while they were in the White House.

- Shortly after the shootings, both Oswald and Booth were
stopped and questioned but allowed to proceed.
- Oswald and Booth were murdered in similar circumstances.
Both were surrounded by their captors under a blaze of
lights. Their assassins, Jack Ruby and Boston Corbett, each
used a Colt revolver and fired a single bullet.

So what accounts for this quite extraordinary catalog of similarities? If you remove errors, distortions, and exaggerations, and make
allowance for the fact that similarities can be found in the lives of any
two human beings, the Kennedy/Lincoln coincidences still take
some explaining.

MORE PRESIDENTIAL COINCIDENCES

In 1992, the *Skeptical Inquirer* held a "Spooky Presidential Coincidences Contest," asking readers to compile lists of coincidences between other pairs of presidents. One of the joint winners, Chris
Fishel, a student at the University of Virginia, came up with many
coincidences relating to Thomas Jefferson and Andrew Jackson. Both
men served two full terms; both their wives died before they became
president; each had six-letter first names; both were in debt at the
time of their deaths; each had a state capital named after him, and
both their predecessors refused to attend their inaugurations.

JULY 4 COMMISERATIONS

In what may have been more than just coincidence, three early U.S.
presidents died on July 4. John Adams and Thomas Jefferson died on

the same day, in 1826, on the fiftieth anniversary of their signing the Declaration of Independence. Adams's final words, that his long-time rival and correspondent Jefferson "still lives," were mistaken, as Jefferson had died earlier that same day. James Monroe died on the same date five years later. Historians suggest that Adams and Jefferson made an effort to hang on till July 4. James Madison rejected stimulants that might have prolonged his life, and he died six days earlier on June 28, 1836. Only one president, Calvin Coolidge, was born on July 4.

IDENTICAL TWINS

Twin boys, born in Ohio, were adopted by different families shortly after birth. In 1979, after thirty-nine years apart, they were reunited. It was discovered that each had been named James; that each had had law-enforcement training; that each liked mechanical drawing and carpentry. Each married a woman named Linda and had a son—one named James Alan and the other James Allan. Each had divorced, and then married a second wife, named Betty. Both had had dogs named Toy. Also, both favored the same St. Petersburg, Florida, vacation beach.

A SINGLE LIFE

Bachelor twins, Bill and John Bloomfield, died as they had lived—inseparably. They lived together throughout their lives, dressed alike, wore the same kind of glasses and kept their hair cut in the same short style. As they grew older, they both had hip replacement operations and took to carrying identical walking sticks. In May 1996 the twins, then aged sixty-one, attended a body building competition. Suddenly one of them collapsed. Officials called an ambulance. The call was logged at 12:14 a.m. At 12:16 a.m. the emergency phone rang again. The other twin had collapsed. Neither man recovered.

THE GIGGLE TWINS

They are known as the "Giggle Twins," but it's not just their identical ringing laughter that has led scientists to study the lives of fifty-eight year olds Barbara and Daphne Goodship.

The twins were separated at birth and had no contact until forty years later when Barbara set out to find her real mother . . . and discovered that she was a twin. "When I met Daphne," says Barbara, "we didn't hug and we didn't kiss. There was no need. It was like meeting an old friend. We just walked off chatting together. The funny thing is we were both wearing a beige dress and brown jacket."

Their identical taste in clothes turned out to be just the first of a remarkable sequence of coincidences linking the twins.

Both met their husbands at a Christmas town hall dance. Both wore blue wedding dresses covered in white lace. Both their first pregnancies ended in a miscarriage in the same month of the same year. Both went on to have two sons and a daughter. Both had their second son in the same month of the same year.

The list of similarities continues. Both, oddly, drink their coffee black and cold with no sugar, both hate heights, both tint their graying hair the same shade of auburn.

At school both hated games and mathematics and read exactly the same books. "Now Daphne rings me and says I've just bought such-and-such a book to stop me buying it," says Barbara.

Geneticists in America, fascinated by the extraordinary coincidences, have been running a variety of behavioral tests on Barbara and Daphne. They found remarkable similarities in their responses to a variety of questions. In one test Barbara was asked to write a sentence. Carelessly she wrote "The Cas Sat On The Mat." When Daphne was presented with the same task she wrote precisely the same sentence, right down to the spelling mistake.

"That sort of thing happens all the time," says Barbara. "Once

Daphne rang for a chat when I was cooking and we discovered we were doing the same recipe."

"We aren't surprised about anything anymore," says Daphne. "I always seem to get illnesses first. I then ring Barbara to tell her what she is going to get!"

UNITED IN DEATH

Twins John and Arthur Mowforth came into the world together, and that was how they left. On the evening of May 22, 1975, the brothers, who lived about eighty miles apart, both suffered severe chest pains and were rushed to separate hospitals. The families of both men were unaware of the other's illness. Both twins died from heart attacks shortly after arrival at the hospital.

PENG AND YONG'S PING-PONG MEDALS

Two Chinese competitors in the 1995 Special Olympics World Games were clearly destined to work together. Their parallel lives began with tragedy but resulted in joint Olympic glory at the New Haven, Connecticut, games.

As a baby, Yong Xi had been seriously burned in a fire. It had destroyed his right hand and severely scarred the right side of his face. He was abandoned at Shanghai railroad station at the age of two.

A couple of months later, two-year-old Peng Kai was abandoned at a Shanghai orphanage. She had also lost her right hand in a fire and was left with just a thumb and a stump. It had also damaged her scalp leaving her unable to grow hair. Neither of the children's parents could be found, so they grew up in the same orphanage.

Both developed a love for table tennis, acquiring extraordinary skills despite their handicaps. When Yong serves, his right arm holds the paddle against his body while he throws the ball high with his left hand. Then he grabs the paddle with the left hand and serves with subtle spins. Peng is able to hold the ball between the thumb and stump of her right hand and throw it while her paddle stays in her left hand.

In New Haven, at the age of twenty-one, they played side by side to win the Special Olympic mixed doubles title. Yong also won gold in the men's doubles and silver in the singles competition. Peng won the gold in the women's singles competition.

.12.

GOOD LUCK

Bad luck is rather too easy to come by. All you have to do is break a mirror, walk under a ladder, or spill some salt. Actually it's much easier than that; you can just stand still—bad luck will find you.

Some psychologists argue that misfortune is the natural state of things. The pessimists, they say, have got it right.

But good luck is achievable. It requires a measure of blind faith, a huge amount of energy, and the ability to see the bad things that happen to you as challenges that will help you become a better person.

Or you could just rely on lucky coincidences like these:

LUCK OF THE AUSTRALIANS

Some people have all the luck.

Alec and Vivienne from Freemantle in Western Australia were able to afford a vacation in London after winning half a million dollars in the State Lottery. They had not long arrived when the news came through that they had won again. This time they'd scooped $876,000. The couple said they planned to continue their vacation. Presumably for ever.

Lottery officials said they had beaten odds of 64 million to 1 to win twice in six months.

THE LOST MAPS THAT WANTED TO BE FOUND

The biologist and coincidence researcher Paul Kammerer noticed that coincidences often come in clusters or series. So has every gambler that has ever lived. Professor C. E. Sherman, chairman of the Civil Engineering Department of the Ohio State University at Columbus, made the same discovery when ten years' worth of good luck landed on his head within twelve hours one day in 1909. Sherman wrote an account of this perfect day in his book, *Land of Kingdom Come,* from which these details are taken.

At the time Sherman was locked in the intractable task of compiling a road atlas of Ohio. The problem was that maps of the southwestern counties of the state were unavailable or nonexistent. The U.S. Geological Survey hadn't yet mapped the area and the only charts to be had were old county atlases. These were usually located in the counties themselves and had to be tracked down and retrieved by letter and parcel. Eventually he managed to secure the atlases for most counties, but there remained two, Pike and Highland, for which he had drawn a blank. Despite all his letter writing, Sherman couldn't be sure maps had ever been made of the regions. Without them it would be a huge task to make a proper road survey. Sherman was also missing a good map of the Ohio River.

The only thing for it was to search for data on the ground, homestead by homestead if necessary. Sherman packed his suitcase one Saturday and wearily boarded a train, telling friends not to expect to see him for two weeks. Incredibly, he found everything he needed in twelve hours.

The first stop was Cincinnati where, in the United States Engineering Office, he found an excellent map of the Ohio River. He then took a train to Highland County, but had to wait at Norwood for a connection to Hillsboro. When he mentioned his quest to the ticket agent he was told, "There's an old book like that in the rear room, I think." Together they searched the old dusty stock room and there was the semilegendary *Highland County Atlas*.

Sherman then took a train to Pike and in the short stop for a connection in Chillicothe he strolled up the street to make an unannounced call on an old friend. No sooner had he set off than he saw his friend walking toward him, as though he had arranged to meet him at the station. They had a chat and then Sherman returned to catch his train. As he was boarding he was hailed by a man who had sent him a letter the day before about some matter, who said he could save Sherman some trouble if he would answer his question on the spot.

Sherman knew only two people at Waverly, the Pike county seat. One was a mechanical engineering student, the other a civil engineering student. He had no idea whether either of them would be there, but spotted the mechanical engineer getting out of the carriage in front of his at Waverly. As they walked toward the hotel together, he said he would send the other man round if he was at home. Sherman had just finished his dinner when the civil engineer turned up. He didn't know of any Pike County map, he said, but his father might. "Here he comes now!"

When asked about the map, the father said he thought the county auditor might have one. At that moment they spotted the auditor walking down the street. It was Saturday night but the auditor invited them straight away to his office in the courthouse across the street and there, behind his desk, hung a fine old map of Pike County.

At this point in his account, afraid that an unleavened diet of good luck might strain credulity, Sherman apologized. "Even the smallest incident seemed to fit perfectly into the harmonious whole," he wrote, before casting doubt on his own impartiality, due to his psychological state at the time. "I had for months been on the quest for all the data and when this last, hardest problem began to unravel so easily, it put me in a humor to notice only favoring circumstances."

But there were an awful lot of those circumstances. For example, the Norwood ticket agent hadn't wanted to sell his atlas, but was happy to lend it; the friend he met in Chillicothe was on his way to the station to catch the train out of town after Sherman's; the tracing paper he took with him at random from a pile that morning just fitted the Pike County wall map; and the civil engineer, who could have been anywhere he wanted in the world, was on hand to help him with the tracing. And who would expect to get into a locked courthouse on a Saturday night to find a map that up till that point he had no idea existed?

"I retired that night," wrote Sherman, "with the sensation of having experienced a perfect day."

ON A WING AND A SPARE

A rare biplane owned by *Jonathan Livingston Seagull* author Richard Bach was upended in 1966 while coming in to land in Palmyra, Wisconsin. The pilots were able to put the plane back together but an essential strut had broken irreparably. The plane, a 1929 Detroit-Parks P-2A Speedster, was one of only eight ever built, so the likelihood of getting another part for it seemed hopeless.

But the owner of a nearby hangar, seeing the broken plane, came over and told them he had lots of plane parts in the hangar to which they were welcome. There, in a pile of parts in the hangar, was the strut they needed to make the plane complete.

In his book, *Nothing by Chance*, Richard Bach writes, "The odds against our breaking the biplane in a little town that happened to be home to a man with the forty-year-old part to repair it; the odds that he would be on the scene when the event happened; the odds that we'd push the plane right next to his hangar, within ten feet of the part we needed—the odds were so high that coincidence was a foolish answer."

DOUBLE JEOPARDY

Lift operator Betty Lou Oliver had the most miraculous escape when a B52 bomber crashed into the Empire State Building in thick fog back on July 28, 1945.

At 9:40 a.m., the aircraft ploughed into the seventy-fourth floor of what was then the world's tallest building. Betty was caught in the ensuing fireball that roared up the elevator shaft, and was severely burned.

She was given first aid treatment for her injuries and then put into a second elevator, which appeared not to have been damaged, to be taken down to meet the ambulance waiting at the bottom.

But, unknown to the rescuers, the second elevator had been damaged by the impact of the aircraft. An engine and part of the bomber's undercarriage had fallen down the shaft and weakened the cables.

As the elevator doors closed, rescue workers heard what sounded like a gunshot as the cables snapped. The elevator hurtled down the one thousand feet from the seventy-fifth floor to the basement.

Incredibly, Betty survived. The severed cables hanging beneath the elevator piled up and acted as a coiled spring, which slowed it down. The descent had also been decelerated by trapped air forming an air cushion at the bottom of the shaft.

Betty had to be cut from the mangled wreckage, but by amazing good fortune, she was alive.

HOCKEY CLUE

Police trying to identify the victims of Fred and Rosemary West faced a massive task. The bodies were all severely decomposed and officers were working with a list of more than ten thousand missing girls. Their only hope was to rely on forensic science, dental records, and in one extraordinary instance, sheer good luck.

Professor David Whittaker, who identified all twelve victims, worked closely with police for a year and a half. He said, "Almost every Tuesday I would speak to all the detectives and help keep up morale.

"One night I put up a photo of a set of remains of a girl who had two temporary crowns on her front teeth. Crowns are made of porcelain and usually take some time to make so dentists fit temporary crowns, often made from plastic.

"She probably had the crowns because she had suffered an injury of some sort or damaged her teeth. I told the detectives she had probably been hit in the mouth or fallen from her bike, had these temporary crowns put on, and was then murdered."

One of the women detectives put her hand up and said she had played hockey against a girl who had been hit by a hockey stick and damaged two front teeth and had them fixed.

"It was a 1 in 10,000 chance but I turned to the detective who was leading the case and said it would have to be investigated as it was a possibility," said Professor Whittaker.

The detective's hunch was investigated and it turned out to be the same girl. Another of the Wests' victims had been identified.

Professor Whittaker added, "However much science we have, and we have a lot in forensic dentistry, the thing we really need is luck and that's what happened in this case."

GUARDIAN ANGEL AT FIVE O'CLOCK

Royal Air Force pilot Derek Sharp is convinced that his amazing escapes from death were more than just coincidence.

Derek's uncle, a Second World War pilot, also called Derek Sharp, led a charmed life as well, experiencing many a close aeronautical shave. Derek sometimes wonders if his uncle has become his guardian angel.

Derek Junior's first dice with death happened back in February 1983. While on a flight with a trainee navigator, his RAF Hawk jet collided head-on with a Mallard duck. The bird smashed through the aircraft cockpit and hit Derek full in the face. It knocked his left eye out of its socket, broke bones in his neck, and smashed bones and nerves in his face. Death seemed certain and imminent.

He still can't really explain how he managed to land the plane. He'd blacked out for several minutes, coming round to discover that the engine had stalled. Almost blind, he somehow managed to restart the engine and land the aircraft at a nearby RAF base.

It was the first of a series of almost unbelievable escapes. On one occasion he was piloting a transport aircraft carrying thirty-five thousand pounds of high-explosive shells in Skopje, Macedonia, during the 1992 Balkans conflict, when a lightning bolt hit the plane right on the nose. It sent a huge fireball along the aisle and out of the tail. Somehow it avoided detonating the explosives.

And during the first Gulf War an American patriot missile accidentally locked its sights onto his plane. "At the last minute they realized that it was aiming at a lumbering old RAF jet instead of an Iraqi scud missile and they aborted it," said Derek. "Strangely I wasn't worried at all as it came toward us. I somehow knew that I was going to be all right."

Derek believes his survival has been the result of more than just chance. His uncle Derek had also defied death on many occasions. During the Second World War he'd been learning to fly in a two-seater Stearman plane when the instructor threw the joystick forward, plunging the nose of the plane down and catapulting Derek Sharp senior, who had failed to fasten his seatbelt, out into the sky, hundreds of feet above the ground. He sailed through the air for several seconds before landing, by incredible chance, on the plane's tail. Somehow the instructor managed to land with Sharp still wrapped around the tail.

Two years later, after several more near-death experiences, Uncle Derek's nine lives ran out when his Lancaster bomber was shot down during a raid over Germany.

His nephew believes the similarities between the two are more than just coincidence. "We had the same name, we look alike, were both pilots, and both had a lot of narrow escapes. He died on the night I was conceived. It's nice to think that if there are such things as guardian angels, maybe Uncle Derek was looking after me."

ALL AT SEA

A schoolboy who got into difficulty while swimming in the sea, experienced two extraordinary coincidences in one day. The first was a one-in-a-million bad luck coincidence, the second an amazing stroke of good luck.

Observers spotted that twelve-year-old Chris Whaites, who was practicing on his bodyboard off the coast at Redcar in northeast England, was in trouble and called the emergency services. The lifeboat that was dispatched came across a windsurfer who by chance had got into trouble in the same area. Assuming the windsurfer was the subject of the alert, the boat returned to base after picking him up, unaware that Chris was struggling for his life nearby.

That would have been the end for Chris had not David Cammish, the launching authority for another lifeboat station, farther up the coast, happened to be listening to the rescue at home on his VHF radio. He realized that 999 calls were still being made after the first lifeboat had finished its rescue operation and immediately launched his lifeboat.

The crew found Chris lying facedown in the water. Mr Cammish said, "I would guess he was only two or three minutes from death. The lifeboat helmsman said when they reached him he was gurgling, lying flat on the surface of the water. He was very, very lucky. It was a million-to-one coincidence."

CHURCHILL'S LUCKY CHOICE

The event that made Winston Churchill a celebrity in 1899 at the age of twenty-four was escaping, with characteristic sangfroid and full expectation of success, from a Pretoria prison during the Boer War. If it hadn't been for an unlikely coincidence he'd have spent the rest of the war back in jail.

Churchill was in South Africa working as a special correspondent for the *Morning Post,* and in this role had been accompanying an armed train to Ladysmith when it was ambushed by Boer guerrillas. Churchill was taken prisoner but managed to escape by climbing out of a latrine window and walking straight out of the prison gate.

He jumped a coal train, hiding among the sacks of coal, but when he realized it wasn't going in the direction he wanted he jumped off again. He wandered about aimlessly for a long time undetected, but became increasingly hungry. Eventually he decided he had no alternative but to knock on a door and seek help. He was in Witbank, a Boer town seventy-five miles from Pretoria and still three hundred miles from the British border. His famous good luck held up. Churchill chose to knock on the front door of the only Englishman in the district, John Howard, a coalmine manager, who concealed him and arranged for him to be smuggled out of the country.

FLIGHT OF ANGELS

A heart attack during a remote transatlantic flight might be considered extreme bad luck. It happened to sixty-seven-year-old Dorothy Fletcher on a trip to Florida, but on this occasion good luck swiftly came to the rescue. When the anxious flight attendant called for a doctor, fifteen cardiologists stepped up. They were all on their way to a cardiology conference in Canada.

Dorothy was in the best of hands and the attack was controlled

with the help of an onboard medical kit. The plane was diverted to
North Carolina, where Mrs. Fletcher recovered in the intensive care
unit of a hospital. "The doctors were wonderful," she said later.
"They saved my life. I wish I could thank them but I have no idea
who they were."

GODS OF CHANCE

Gabriel García Márquez reveals in the first volume of his autobiog-
raphy, *Living to Tell the Tale,* how, as a young adolescent, he embarked on
an arduous journey across Columbia to Bogota in the hope of being
allowed to take an examination for a school scholarship. He didn't
rate his chances very highly. En route he happened upon another
traveler who gave him the gift of a book in return for teaching him
how to sing a romantic bolero.

Upon arriving in Bogota he was dismayed to find hundreds of
students in line in the rain at the ministry of education. Joining the
end of the line his spirits were very low.

Feeling a tap on his shoulder, he turned to face the man he had
met on his journey, who asked him what he was doing there. When he
explained, the man laughed and revealed that he was Dr. Adolfo
Gomez Tamara, national director of scholarships for the Ministry of
Education.

Marquez says, "It was the least plausible coincidence, and one of
the most fortunate of my life." Marquez was promptly registered for
the scholarship examination with no further formalities. He adds,
"They told me first that they were not showing contempt for application forms but paying tribute to the unfathomable gods of chance."

OUR REGULAR ROBBER

The secret of success in life, more often than not, is to find a winning formula and stick to it. In the case of bank robbery however, it

is prudent to vary your strategy in order to avoid being caught. Unless that is, you have coincidence on your side.

One who certainly had more than his fair share of luck took a liking to a bank in Detroit. Entering the bank, he walked up to the teller at the first window and passed a note under the window that read: "I have a gun. Give me all your 100s, 50s, and 20s. Don't give me the bait money or the dye pack or you'll be sorry." The teller placed the money in a big brown envelope and gave it to the robber. He picked it up, walked right out past the guard onto the main thoroughfare, and disappeared into the crowd.

As soon as he had left the teller pushed the alarm button. The closed-circuit cameras TV had been running but no images were captured because the recorder had run out of tape two days earlier and no one had replaced it with a new one.

A few weeks later, the same robber entered the same bank and approached the same teller and passed the same note. Once again the teller filled a brown envelope with money and the robber took it and left the bank. There was no guard on the door this time. He and the manager were in the back office reviewing films of the past three weeks in the hope of getting a glimpse of the robber casing the bank on a day when the cameras were recording. Because they were using the equipment, there were again no closed-circuit TV pictures of the robbery.

ELECTRIC PERFORMANCE

During the making of Mel Gibson's biblical epic, *The Passion of Christ*, Jim Caviezel, the actor playing Jesus, was struck by lightning as he hung from the cross. The devout Gibson rejected coincidence as an explanation, preferring, as Caviezel was unharmed, to see the incident as a sign of heavenly endorsement. It seems a funny way to say good job, though it's certainly true that the fundamentalist God Gibson recognizes would have fried the actor had He been displeased. A skeptic would point out that it can be dangerous hanging from crosses on hilltops in inclement weather.

.13.

DATES, NUMBERS, AND WRONG NUMBERS

Our lives are full of numbers—from addresses to telephone numbers and bank account numbers to house alarm codes. We have the ability to remember a vast amount of digits and the capacity to spot when the numbers emerge in some surprising and coincidental ways.

For example, the address of Howard Trent of Fresno, California, ends with the digits 742, as do his telephone number and bank account number. The number of a compensation check he received after an injury was 99742, which matched the last five digits of his telephone number. The serial numbers of a set of new tires he bought ended in 742 and the number of his car license plate is FDC742.

Some numbers are loaded with cultural significance. Our birth date is particularly special to many of us, tied as it is to beliefs that our entire fate is determined by it. Other numbers are thought to relate to bad luck or danger. The number 666, for example, is considered the "mark of the beast" and for millions of people the number 13 is a certain harbinger of ill-fortune.

Engineers working on India's Hassan-Mangalore railway line back in the late 1970s may well have suffered from triskaidekaphobia—fear of the number 13. They reported major problems with the construction of tunnel number 13. A series of five major rockfalls held up work for months. According to the *Rail Gazette International* of April 1979, "The tunnel was renamed No. 12-A and suddenly all was well."

Astronauts are no greater fans of the number 13, since the explosion of an oxygen tank prevented the ill-fated *Apollo 13* from reaching the moon, and almost cost the lives of its crew.

Nature is full of numerical coincidences. Mathematician Ian Stewart points out that many flowers have five or eight petals, but very few have six or seven. As seemingly random a thing as a snowflake always has "six-fold symmetry." Our entire universe is full of mathematical coincidences—most of them not yet fully understood.

How to Lose Seven Shillings

The following memoir, sent to Arthur Koestler after the publication of his book *The Roots of Coincidence* in 1973, ought perhaps to be in the apocrypha section of this book.

The author of the letter, Anthony S. Clancy of Dublin, Ireland, writes, "I was born on the seventh day of the week, seventh day of the month, seventh month of the year, seventh year of the century. I was the seventh child of a seventh child, and I have seven brothers; that makes seven sevens. On my twenty-seventh birthday, at a horse-race, when I looked at the racecard to pick a winner in the seventh race, the horse numbered seven was called Seventh Heaven, with a handicap of seven stone (ninety-eight pounds). The odds were seven to one. I put seven shillings on this horse. It finished seventh."

STOCK MARKET SCAM

The rise and fall of stock market prices is notoriously difficult to predict. Playing the market can be a fast road to penury, so when one particular stockbroker started to demonstrate an almost superhuman capacity to detect market trends, he found his prediction services in great demand. Was it down to pure chance, coincidence—or something else?

In fact, in his particular case, it was something beyond coincidence . . . though nothing of a paranormal or supernatural order. This publisher of a stock newsletter would send out sixty-four thousand letters extolling his state-of-the-art database, his inside contacts, and his sophisticated econometric models. In thirty-two thousand of these letters he would predict a rise in some stock index for the following week—and in the other thirty-two thousand he'd predict a decline.

Whatever happened to the stock market that week, he'd send a follow-up letter—but only to those thirty-two thousand people to whom he'd made a correct "prediction." To sixteen thousand of these he'd predict a rise for the next week and to sixteen thousand a decline. Whatever happened he would have sent two consecutive correct predictions to sixteen thousand people. And so on. In this way he built the illusion that he knew what he was talking about.

His purpose was to boil the database down to the 1,000 people who had received six straight correct predictions (by coincidence) in a row. These would think they had a good reason to cough up the $1,000 the newsletter publisher requested for further "oracular" tip-offs.

DOPPLEBANGER

When Ernest Halton parked his car outside his church he noticed that the car next to his was the same make and color. That was unusual but not extraordinary. However the next thing he noticed was incredible: the car had the same license plate number. Halton asked among the church congregation for the owner, who turned out to be Tony Gowers, a man he knew, who was as surprised as he had been. Gowers had bought the car second-hand four weeks before. Gowers's car's license number was actually one numeral different to Halton's, but Gower had reconditioned the car when he bought it, and as part of the process had ordered fresh plates. A slipup at the plate makers resulted in one of the numerals being printed wrongly.

WRONG NUMBER, RIGHT CHOICE

Like many teenagers who've fought with their parents, Julia Tant walked out of her parents' house in a fury, vowing never to return, with her suitcase in her hand. She went first to her local youth club to cool off and there a good friend persuaded her to phone her mother, if only to let her know where she was going.

In her agitation, without realizing it, she dialed the wrong number. A woman who sounded like her mother answered. "It's me," Julia said.

"Where are you?" said the woman.

Julia told her she was at the youth club and that she was going to her grandmother's. At this point the woman started swearing at her and shouting, "Julia, get home here!"

Despite the fact that she had used her name, Julia was beginning to realize that something wasn't quite right. "Why are you swearing?" she said. "You never swear."

By this time the woman too was realizing this Julia was not her daughter. She composed herself, explaining that her daughter Julia had walked out on her and disappeared.

Julia had stumbled on a situation more serious and extreme than her own, yet alarmingly similar in many respects. The telephone call sobered her and after it she returned home to be reunited with her parents. Now, years later, she says, "I felt it was a kind of omen so I went back home. If it wasn't an omen it certainly seemed like one."

LAST PUTT

In December 1991 golfer Tony Wright died on the fourteenth green of his local golf course, fourteen months after his father Les collapsed and died at the same spot. Both men had been lining up for putts when they suffered heart attacks.

LAST SYMPHONY

Beethoven, Schubert, Dvořák, and Vaughan Williams (among others) all died after composing a ninth symphony. Mahler, superstitious about his ninth, urgently commenced his tenth as soon as he'd finished his ninth, but not quickly enough to avert his death. Bruckner went to elaborate lengths to delay the syndrome, numbering his first two symphonies 00 and 0. To no avail: he died while composing his ninth. Sibelius stopped after his eighth and lived another thirty-three years.

THE BIRTHDAY CARD THAT NEVER GAVE UP

When Mrs. J. Robinson's mother died in 1989, she found among her possessions a birthday card she had sent to her niece in 1929. The card had been returned by the post office because her mother had got the address wrong. Though they were not in regular contact, Mrs. Robinson thought her cousin might still like to see the card, so she posted it to her again. She had no idea that it would arrive on her birthday, exactly sixty years late.

UNDER COVER

A picture Mrs. G. L. Kilsby inherited from her mother had started to get on her nerves, so on March 1, 1981, she decided to open it up. Inside was the cover of a magazine whose date—March 1, 1881— showed it to be exactly one hundred years old to the day. Mrs. Kilsby was so pleased with the coincidence that she threw away the picture and framed the magazine cover.

FATE OF BIRTH

Baby Emily Beard came into the world on the twelfth day of the twelfth month at twelve minutes past twelve. Her dad David was born on the fourth of the fourth at forty minutes past four. Mother Helen was born on the tenth of the tenth. Brother Harry, three, was born on the sixth of the sixth. Grandmother Sylvia Carpenter was born on the eleventh of the eleventh.

Emily nearly ruined the pattern. She was due two hours earlier but complications put the birth back. David, from Gosport, Hampshire, said. "It's quite spooky; like 'You've entered the twilight zone.' It was only when I rang my mom to tell her about Emily that she told me I'd been born at 4:40. That's when we realized just how weird it all was."

LUCKY FOR SOME

The novelist David Ambrose has been dancing a tango with the number thirteen.

His uncanny association with the number considered by many to be unlucky began when he was writing a novel called *Superstition*. Working at a computer, he would check every day or so how many words he had written. The word-checker facility would also tell him how many lines and paragraphs he had written and the average number of words per sentence.

"I found that I was consistently writing an average of thirteen words per sentence," says David. "I thought maybe I always wrote thirteen words per sentence; I found it hard to believe I was doing it only now, unconsciously, because I was writing a novel called *Superstition*. But when I checked the manuscripts of other novels and stories of mine, I found my average was fourteen to sixteen words per sentence, never thirteen."

Much earlier, before he had begun properly writing the novel, he had sold the film rights on the strength of a thirteen-page outline. "I do not recall consciously registering the moment at which the deal was struck," he says. "However, I saw in my diary shortly afterward that it had happened on the afternoon of Tuesday the thirteenth of February, 1996. The producers of the film asked me to meet them at the Cannes Festival in 1997. The only day we could all manage turned out to be Tuesday the thirteenth of May. In June they flew me out to L.A. for further meetings. Still nobody was actively thinking 'thirteen.' I arrived on the eighth, planning to fly on to New York to see the American publishers of *Superstition* the following Friday— which turned out to be the thirteenth of June. While in L.A. I picked up from my agent a copy of the fully executed contract. The covering letter from the agency's legal department was dated the thirteenth of May—coincidentally the same day on which I had lunched with the producers in Cannes."

Despite every effort to finish earlier, he eventually completed the screenplay version of *Superstition* on October 13. "In February 1998 I had to have a meeting with my London publishers to discuss the paperback edition of the book. The only date on which it turned out that everyone could be there was Friday the thirteenth of February."

David stresses that he had been totally unaware of all these coincidences at the time. "If we had been making this happen even half consciously, we would surely have published the book on the thirteenth," he says. "As it was, it came out on the tenth of July. However, my editor and I weren't able to have our planned celebratory lunch on that day, so it had to be moved to the following Monday. Which was the thirteenth."

The final coincidence related to the number of radio and TV shows at which David was interviewed about the book. "Nobody gave any thought to the actual number of shows," he says. "It was, as always, just a question of getting on as many as possible. At the end of the week, when I glanced through my schedule and counted up the number of broadcasts I'd done, I saw it was thirteen."

TWIN NUMBERS

The winning number in the evening drawing of the New York Lottery three-digit "numbers" game on September 11, 2002 was 911.

CAPTAIN CLARK'S BAD DAY

The novelist William Burroughs had a thing about the number twenty-three. It kept on cropping up in the coincidences he noticed. He met a Captain Clark, when living in Tangier, in 1958, who boasted he had been sailing twenty-three years without an accident. That day Captain Clark took his boat out and had an accident. Later Burroughs heard a news report on the radio about an airline crash. The flight number was twenty-three; the pilot's name was Captain Clark.

BIBLE NUMBERS

Keen-eyed numerologists have spotted some interesting coincidences in the Bible.

They point out that Psalm 118 is the middle chapter of the entire Bible; that, just before it, Psalm 117 is the shortest chapter in the Bible and Psalm 119 is the longest chapter.

The Bible has 594 chapters before Psalm 118 and 594 chapters after Psalm 118. If you add up all the chapters except Psalm 118 you get 1,188 chapters. If you take the number 1,188 and interpret it as chapter 118 of Psalms, verse 8, you will find the middle verse of the entire Bible—"It is better to trust in the Lord than to put confidence in man."

Some would say that this is the central message of the Bible—numerically speaking at least.

PASTOR BLASTER

Alicia Keys couldn't have been more inviting on her hit single, "Diary." "Ooooh baby," the singer crooned, "if there's anything that you fear . . . call 489-4608, and I'll be here." The trouble was, she wasn't there. Fans in Georgia who dialed the number got an increasingly exasperated retired pastor instead.

While the song was high in the R&B/Hip-Hop charts, J.D. Turner, who happened to have the same telephone number, was getting twenty to twenty-five calls a day at his Statesboro home. Sometimes they'd drag the poor man out of bed. "They call at 4:30 a.m.," he said, "and then say, 'I want to talk to Alicia Keys.'"

The number in the song is genuine. It used to be Keys's own number when she lived in New York. Used with the 347 area code it connects to an answering service where fans can record a message. But used with a 912 area code it merely means extra nuisance for Mr. Turner in Statesboro.

"I don't want to change my number," said Turner, who would face a lifetime of sleepless nights if the song achieves all-time-great status. "I've had the same number for fourteen years."

CATCH A FALLING STARFISH

In 1996 an Englishwoman named Ellen was vacationing on the north coast of France with two friends. She had heard that starfish were often to be found on a local beach and hoping to see one, spent a morning walking the beach with her friends. But it was a disappointing outing. It was cold, wet, and windy and not a single starfish was to be seen.

Eventually the three gave up and drove to the coastal town of Calais. They parked their car in the main square and as they got out Ellen remarked what a shame it was she hadn't found a starfish. At that moment, as though her words were a signal, a single starfish fell from the sky and landed at her feet. Unseen by Ellen, but spotted by her friends, two seagulls had been fighting over the starfish above their heads and dropped it in the squabble.

DO THE OKLAHOMA NUMBERS ADD UP?

Great tragic events linger long in the public imagination. The facts are dissected, analyzed, and subjected to such a concentration of sometimes lurid imagination that extraordinary theories are gleaned from them.

The so-called "significance" of the dates of the bombing of the Murrah Building, Oklahoma, in 1995 and the subsequent execution of the perpetrator, Timothy McVeigh, is an example of numerology, in which sensationalists have selectively manipulated numbers to promote, in this case, a sinister mystic resonance. Consider the following figures:

04—The month of the Oklahoma City Bombing (April 19, 1995)
19—The day

95—The year
09—The hour the bomb went off
02—The minute
06—The month McVeigh was executed
11—The day
01—The year
07—The hour he was pronounced dead
14—The minute

The total, 168 equals the number of people killed.

Is this coincidence or some dark force at work? Discounting the dark force that motivates a human mind to seek such naive significance in distressing events, it is certainly coincidence. This is the same thinking that lies behind the Bible codes: take all the many numbers relating to an event—dates, casualties, distances traveled, etc.—fiddle around with them for days (adding, subtracting, multiplying, matching, eliminating...) until you hit on a pattern, then pronounce the result meaningful. As with all such meaningful combinations, the meaning itself cannot be discerned.

DOUGLAS ADAMS WAS (ALMOST) RIGHT

Scientists at Cambridge University were amazed when, after three years calculating one of the fundamental keys to the universe—The Hubble Constant—they came up with the number forty-two. This is the same number that a computer called Deep Thought in Douglas Adams's famously hilarious science fiction novel *The Hitchhiker's Guide to the Galaxy*, decides after seven million years of calculations, is the answer to life, the universe, and everything.

In Adams's hands, of course, the figure sounded suitably absurd, but coming from august Cambridge scientists it has super-enhanced gravitas. The Hubble Constant is the velocity at which a typical galaxy is receding from Earth, divided by its distance from Earth.

This is no trivial concern, for the figure determines the age of the universe. It should be added, for those who think this is all too easy, that the number refers to kilometers per second per megaparsec (a unit of distance used in astronomy).

It would have been an extraordinary coincidence, but alas, it was too good to be true, for the figure is hotly debated. Recent estimates range between fifty and one hundred, showing those original scientists to be off the mark. Or was it just wishful thinking on their part?

"It caused quite a few laughs when we arrived at the figure forty-two," said Dr. Keith Grange, "because we're all great fans of *The Hitchhiker's Guide*."

. 1 4 .

PSYCHIC
COINCIDENCES?

If you dismiss the possibility that a great many, if not all, of the stories in this book are the result of simple coincidence, then you have to look for another explanation.

As we have learned, great minds such as those of Arthur Koestler, Wolfgang Pauli, and Carl Jung have attempted to find evidence, theoretical or otherwise, of some sort of universal unifying force that explains the kind of phenomena that are so often dismissed as pure chance.

In his book *Incredible Coincidence* Alan Vaughan writes:

> *I dreamed I was talking with the parapsychologist Gertrude Schmeidler about synchronicity. She asked, "But where does synchronicity end and chance begin?"*
>
> *"But don't you see," I exclaimed. "Everything is synchronicity. Nothing happens by chance." As I said these words in my dream, a tremendous energy*

flooded my brain and shocked me awake, forcing me to consider this intuitive answer.

What if it were true? What if, moment by moment, we create our own realities through our consciousness? Literally.

Well, that might explain some of the following stories.

DREAM WOMAN

Pat Swain was on her honeymoon in Bled, Slovenia, when she dreamed she saw her best friend's cousin walking with her sister. In the dream Pat was looking out of a window and looked down on the two women below. The odd thing about her dream was why it had featured those particular women, as she hadn't seen Hilda, her friend's cousin, for twenty-five years, and the sister, Stella, she barely knew. They were certainly not on her mind at the time.

Two days later Pat and her husband visited the castle on the cliffs above Bled. She was looking out over a wall when she spotted the two women of her dream walking along the tier below. They went down to greet them and Hilda told her they had come to Bled on a day trip from their resort on the coast. "A split second either way and we would have missed one another," said Pat.

PAIN TRAVELS

Susie Court dreamed her friend Elaine Hudson was in terrible pain in a hospital bed. She hadn't spoken to her for months but this dream was very vivid. Her friend was writhing in agony in the dream and Susie tried to comfort her. It was the night of August 1, 1988.

Susie woke up worried for her friend and tried to phone her. But Elaine had moved and her number was unlisted. It took her several hours phoning mutual friends before she managed to track down Elaine's sister-in-law. Was Elaine okay? she wanted to know.

Very well, said the sister-in-law. Elaine had given birth to a little boy, Sean, during the night. "I didn't even know she was pregnant," said Susie.

CALLING ESMERELDA

The ring of a telephone very early in the morning woke Mrs. M. Rigby while she was staying at a friend's apartment in 1976. Realizing that it wouldn't be for her she went back to sleep and fell into a dream of telephone calls. She dreamed she was in the same bed, in the same apartment, but that when the telephone rang again she got up and went to the hall to answer it. The mournful voice of a woman asked if she could speak to Esmerelda. Mrs. Rigby said she was sorry, but there was nobody of that name in the apartment. Again the woman asked, and again Mrs. Rigby told her there was no Esmerelda there. In the dream she hung up and got back into bed.

Later that morning, over breakfast in the kitchen, she asked her friends who had called so early. One of the friends, Johnny, said that it had been his mother. Mrs. Rigby then described her dream and Johnny went quiet. He said that before he was born his mother had given birth to a little girl who only lived three weeks. Before she died they had her baptized. Her name was Esmerelda.

ATTACK OF WIND

Nineteenth-century French occultist and astronomer Camille Flammarion was writing a chapter about wind in his book on the atmosphere, when a gust blew through his window, lifted the loose pages he'd just written from his desk, and sucked them back out of the window. Days later he received a routine parcel of his latest proofs from his publisher containing transcripts of the very pages that had gone missing. The porter, who acted as a regular messenger for Flammarion, solved the mystery. He had been passing the house by chance, saw the pages in the street, collected them up and took them to the publisher in the normal way.

DOG DAYS IN COMORO

Ali Soilih was a tin-pot dictator with a superstitious streak. Four weeks after the Comoro Islands, situated between Madagascar and the African mainland, gained independence from France in 1975, Soilih seized power with the help of French mercenary Colonel Bob Denard, and subjected the islanders to a tyrannical regime. A witch told him that he would meet his end at the hands of a man with a dog, so Soilih had all the dogs on the islands put to death. That was when Denard, now working for the other side, arrived to confront his former boss. He was leading an Alsatian. Whether Denard knew about the prophecy, which seems likely, and brought a dog with him deliberately to sow ominous fears, we don't know.

THE LITTLE BOOK

In *The Challenge of Chance*, Arthur Koestler tells of a synchronistic episode concerning a book—an experience so dramatic that it converted him to a belief in psychic phenomena.

The year was 1937 and Koestler was imprisoned in Spain by the Franco regime awaiting the order for his execution.

"In such situations," writes Koestler, "one tends to look for metaphysical comforts, and one day I suddenly remembered a certain episode in Thomas Mann's novel *Buddenbrooks*. One of the characters, Consul Thomas Buddenbrook, though only in his forties, knows he is about to die. He was never given to religious speculations but now falls under the spell of a 'little book' that for years had stood unread in his library, in which it is explained that death is not final, merely a transition to another, impersonal kind of existence, a reunion with a state of cosmic oneness. The book was Schopenhauer's essay on death."

Koestler was exchanged for a hostage held by the other side and the day after his release he wrote to Thomas Mann to thank him for the comfort he had received from the passage reflecting on Schopenhauer's essay. Mann replied that he had not read the essay for forty years, but a few minutes before the postman handed him Koestler's letter he had had a sudden impulse to fetch the volume from his library.

JUNG STRIKES BACK

The writer and adventurer, Laurens van der Post, tells this story about his friend Carl Jung, in his book *Carl Jung and the Story of Our Time.*

"I was making a film of the story of the life of Jung some years ago. The time schedule for the film had been determined nearly a year before we started filming. The final sequence on the last day of all was to be filmed in Jung's old house. We had worked all morning in his home and all day long the cameraman, producer, and myself— without mentioning it to one another—had an indescribable feeling that Jung was near to us. I heard the cameraman saying to an assistant, half jokingly at the time, 'You know I had a feeling as if Jung were looking over my shoulder all the time.'

"It was a dry, hot, blazing afternoon and we left the house at lunchtime to do some background filming in the afternoon in the oldest section of Zurich, intending to return for the filming of the final scene by his home at sunset. On our way from Zurich to Kusnacht to do so, suddenly out of the hot blue sky the thunder clouds tumbled without forewarning, as if in a great hurry. By the time we reached Kusnacht the lightning was flashing, the thunder rumbling, and the rain pouring down.

"When the moment came for me to speak direct to the camera about Jung's death and I came to the description of how the lightning demolished Jung's favorite tree (two hours after his death) the lightning struck in the garden again. The thunder crashed out so loud that I winced and to this day the thunder, wince, and the impediment of speech it caused are there in the film for all to see."

BOUNCING BABIES
AND GOLF BALLS

Some coincidence stories defy classification. . . .

THE INSTANT GRANDAD

Texan Ron Thompson's family grew by four in less than twenty-four hours in 1990, when three of his daughters, Mary, Joan, and Carol, gave birth to four boys.

First in line was Mary, aged twenty-eight, who was driven to the hospital by her nineteen-year-old sister Joan (also nine months pregnant). Five hours later, Mary gave birth to Shane. Seven hours after that, Joan was herself driven to hospital, by her pregnant sister Carol, and gave birth to a boy, Jeremy, a minute after midnight. Then twenty-four-year-old Carol went into labor, delivering twin boys just before 3 a.m.

Hole in One

American golfer Scott Palmer claims to have hit nineteen holes in one. The chances of getting one are about 43,000 to 1. Scott, who has rounded up sixty-five witnesses to verify his claims, says he hit four of them on consecutive days in October 1983.

Palmer says his method is to conjure up a mental image of a faceless woman pouring a glass of milk at the moment he hits the drive. Sounds simple enough.

Death Takes a Holiday

In 1946 Mildred West decided to take a week's vacation. She was the obituary writer of the *Alton Evening Telegraph*, New York. For the first time in the memory of anyone on the newspaper, during the seven days she was away, there were no deaths recorded in Alton, a city of thirty-two thousand. Normally they averaged ten a week.

One Giant Home Run for Mankind

San Fransisco Giants' baseball pitcher Gaylord Perry recalls how his former manager once rather pessimistically predicted that "they'll put a man on the moon before he hits a home run."

The unflattering assessment, made by manager Alvin Dark in 1964, was a bit unfair, as Perry had a fairly respectable batting average for a pitcher, but it turned out to be uncannily accurate.

Six years later, on July 20, 1969, during a home game against the Dodgers, Gaylord Perry finally hit his first home run—but he was nipped at the wire by *Apollo II*'s lunar module that had touched down on the moon just minutes earlier.

NOISY NEIGHBORS

Two commemorative blue plaques in a London street reveal that Jimi Hendrix and George Frederick Handel lived next door to each other.

Handel (1685–1759) lived and died at number 25 Brook Street; Hendrix (1942–70) lived for one year at number 23.

MONK TO THE RESCUE

The nineteenth-century Austrian portrait painter Joseph Aigner had a death wish, but thanks to repeated interventions by a Capuchin monk, it took him fifty years to realize his ambition. He first attempted to kill himself when he was only eighteen. His clumsy efforts to hang himself were interrupted by the arrival of the mysterious monk. Four years later he made a second attempt to hang himself, but was again thwarted by the same monk. At the age of thirty it looked like his death wish would be fulfilled when he was sentenced to be hanged for his political activities. Once again he was saved by the intervention of the monk.

Aigner was sixty-eight before he finally succeeded in ending his life. He shot himself with a pistol. His funeral ceremony was conducted by the same Capuchin monk—a man whose name Aigner never even knew.

THE FOUR TOWERS

In the wake of the World Trade Center tragedy on September 11, 2001, a group calling itself The Two Towers Protest Organization began a campaign to prevent the second film of the *Lord of the Rings* trilogy being named *The Two Towers*. Although the film is named after J. R. R. Tolkien's book, written and titled more than fifty years ago, the protest organization rejected the argument that the title was merely an innocent coincidence.

The organization, which described itself as being made up of "like-minded individuals who were greatly affected" by 9/11, issued a statement that read: "We believe that Peter Jackson [the film's producer] and New Line Cinema's actions are in fact hate speech. The movie is intentionally being named *The Two Towers* in order to capitalize on the tragedy of September 11. Clearly, you cannot deny the fact that this falls under hate speech. We believe that if they will not willingly change the name, the government should step in to stop the movie's production or to force a name change."

The *Two Towers* film project was named long before the tragedy in keeping with the second book of Tolkien's trilogy. Immediately after the September 11 crisis Jackson did briefly consider renaming it, but decided against it because it would upset Tolkien enthusiasts, and also the book bearing the name is permanently in mass publication. Another fact that everybody seemed to overlook in the controversy is that the World Trade Center was never referred to as the Two Towers, it was always the Twin Towers.

THE UNFORTUNATE ANAGRAM

Naturalist Sir Peter Scott was an enthusiastic believer in the Loch Ness monster. So great, in fact, was his confidence in the creature's existence that he promoted the use of the Greek name for it: *Nessiteras rhombopteryx*. This name, which he and underwater photographer Robert Rines coined in December 1975, may be roughly translated as "the Ness monster with diamond-shaped fin." As London newspapers quickly pointed out, the name is also an anagram for the words "Monster Hoax by Sir Peter S."

HERE A MOO

The postal code of a Canadian farmer called MacDonald contained the letter sequence EIEIO.

POOR TOSSER

In an attempt to demonstrate a fifty-fifty probability in his first lecture at a new university, a professor of statistics tossed a coin that landed on a smooth floor on its edge. The likelihood of this happening has been estimated at approximately 1 billion to 1.

DUDLEY DANGER

The comedian Peter Cook once wrote about how his "minute seaweed-eating partner" Dudley Moore had an irrational fear about one of their more surreal sketches. The particular routine involved a graphic, if somewhat scatological, account of lobsters crawling up the bottom of the late actress Jayne Mansfield.

Said Cook, "Dudley was terrified of being beat up by Mickey Hargitay, Jayne's muscular ex-husband. I talked to Dudley yesterday. He has just rented a house in Los Angeles for six months. Only after he moved in did he discover the identity of his next-door neighbor. Yep, Mickey Hargitay."

A TOWN LIKE ALICE

In a man's world, the town of Pacifica, California, was a remarkable exception.

Back in 1992, the thirty-eight thousand residents of the town, which lies just to the south of San Francisco, voted in the first all-female city council in California and the first in the country for more than a century.

"I earnestly feel with all my heart and soul that it's not a woman's issue in Pacifica," said Barbara Carr, a real estate agent who was one of four women voted onto the council. "It's just coincidence."

The extraordinary turn of events came about after the three remaining men on the council, and one woman, were ousted in a recall election, held following a battle over a lighting and landscaping tax.

No one expected all seventeen men running for the council to be defeated in the subsequent election, and four of the five women in the race to come out on top.

"I really don't think the voters made any conscious decision to elect only women," said councilwoman Bonnie Wells, who is likely to be chosen by the others as the new mayor. "That's just the way results turned out. I think they chose four people who could do the job and they happened to all be women."

. 16 .

APOCRYPHA

It is not possible to guarantee the absolute authenticity of every story in this book. Coincidence stories are often exaggerated, distorted and—God help us—invented.

For example, coincidence fans take great delight in pointing out that there was a tornado in Kansas the day Judy Garland died.

Well that's not such a remarkable thing. Kansas does have the third highest incidence of tornadoes of all the states in America, sometimes experiencing more than one hundred a year. As it happens, tornadoes were comparatively rare in 1969—only seventeen of them. But that still makes it quite likely that a tornado might have come along to blow the spirit of Judy Garland somewhere over the rainbow.

Such is our love of coincidence that we often feel compelled to embellish the facts. We can't help ourselves. But it does make it very difficult to distinguish between true stories of incredible coincidence and those that are a touch apocryphal. Take, for example, the many accounts in circulation of precious items that turn up in the stomachs of fish. Stories like these:

> *Norwegian fisherman, Waldemar Andersen, made an amazing discovery as he gutted a cod he had just caught in the North Sea. Inside its stomach was the gold earring his wife had lost the previous week.*

Here's another version:

> *In the summer of 1979, fifteen-year-old Robert Johansen caught a ten-pound cod in a Norwegian fjord. He gave it to his mother to prepare for dinner. Inside its stomach she found a diamond ring—a family heirloom she had lost while fishing in the fjord ten years ago.*

And another:

> *Joseph Cross of Newport News, Virginia, lost his ring in floodwaters during a storm in 1980. In February 1982 a restaurateur in Charlottesville, Virginia, found the ring—inside a fish.*

The original ring in the fish's stomach story probably goes back to the days of ancient Egypt.

Herodotus tells how the Pharaoh Amasis advised his ally Polykrates to throw a precious ring into the sea as a gift to the Gods. Polykrates did this, but a few days later a fisherman brought him a gift of a large fish he had just caught. Inside its stomach was the ring. It was a bad omen. Hearing what had happened, Amasis ended their alliance and shortly afterward Polykrates was murdered.

That ancient tale sounds more likely than this story reported, as true, in *Ripley's Giant Book of Believe It or Not*:

> *The wife of Howard Ramage lost her wedding ring in a drain in 1918 and a Vancouver man found it thirty-six years later in the stomach of a fish and returned it to Mr. Ramage.*

These stories crop up all over the place—sometimes from fairly respectable sources. The stomach involved is not always that of a fish:

> On March 28, 1982, both the *Sunday Express* and the *News of the World* reported that two years after farmer Ferdi Parker lost an antique wedding ring, a vet found it in a cow's stomach while performing an autopsy.

There's a clear lesson here: if you lose your ring, look in the stomachs of animals. Here's another example:

> Evelyn Noestmo lost her wedding ring in 1993, pushing her car out of a ditch. It turned up in the stomach of a moose shot by her husband in 1996.

And how likely is this?

> An American fisherman severed his thumb in a boating accident. The missing digit was eventually recovered from the stomach of a fish.

Doesn't sound very likely, does it? But according to coincidence expert Ken Anderson, it is quite true. Anderson reports that the accident-prone fisherman was tracked down and the "facts" verified beyond dispute.

The man was thirty-two-year-old American, Robert Lindsey, a welder with the Union Pacific railroad. He'd gone out fishing on a friend's boat at the Flaming Gorge Reservoir in Wyoming when the wake from passing cabin cruisers washed him overboard. He was swept under the boat where the propeller blades severely cut his leg, severed his thumb and nearly removed two fingers.

Some six months later, after he had more or less recovered from his injuries, his wife showed him an article in a newspaper. Lindsey takes up the story himself, "The article was about a man out fishing who'd found a human thumb inside a fish he had caught, five miles from where I had my accident. I figured it was probably my thumb so

I called up the coroner, who thought I was joking at first. But then I went to see him and the thumb. It looked like my thumb. It was in good shape considering."

Further X-ray tests confirmed it was Robert's thumb. He now keeps it in a jar of formaldehyde.

One of the more truth-resistant of apocryphal stories insists that Clint Eastwood is the son of Stan Laurel. The story derives from a coincidental facial similarity in the two men. Probably nobody would have ever noticed this had not an Italian newspaper pointed it out, printing adjacent portrait photographs of the two smiling men. Then a British children's magazine published a photograph of Eastwood with his hair sticking up on end, suggesting he could be Laurel's son. The rumor spread like wildfire, aided by Eastwood's notorious reticence to discuss details of his private life, which made reliable information about him rarer than unicorns. Then there is the second coincidence that Eastwood was born on May 31, 1930, the same year and month that Stan Laurel's wife Lois gave birth to a son, who died nine days later. The most imaginative urban-mythtellers like to claim that the baby in fact survived and went on to an illustrious career in Spaghetti Westerns, totally ignoring the fact that there is no mystery about Clint Eastwood's parentage at all. He was born in San Francisco, of Clinton Eastwood Sr. and Margaret Eastwood.

One oh-my-God story circulating widely since the al-Qaeda train bombings in Madrid asserts that there were exactly 911 days between the 9/11 attack on the World Trade Center in New York and the Madrid bombs on March 11, 2004. Well there would be exactly 911 days if it wasn't for the fact that there are exactly 912 days. A miscalculation perhaps? Forgot the 2004 leap year? No, say proponents of this creepy but meaningless assertion, all you have to do to arrive at the exact number is to not count March 11 itself. Ah, almost exactly 911.

We leave you with a selection of stories that have the glitter of make-believe about them. That, of course—as we now know—isn't proof that they couldn't, or didn't, occur. Are they fact or fiction, incredible coincidence or something beyond coincidence? You decide.

WRITING WRONGS

A handwriting expert offered to do readings from samples sent to her. A woman sent a note written by her boyfriend and asked if he would make a good husband. The graphologist replied that this was very unlikely as he had been "a terrible husband to me for the past three years." And in a postscript she thanked her for the "evidence."

LOVER'S LEAP

After discovering that her husband had been unfaithful, Vera Czermak of Prague threw herself from her third-floor balcony, only to land, by extraordinary chance, on her husband as he walked directly below. She killed him, but escaped with minor injuries herself.

THE WRONG TAXI

An Athenian taxi driver was astonished when the customer he picked up gave his own address as his destination. Arriving at his home, the driver watched as the man took out a key and let himself in. Using his own key the driver followed and interrupted the man as he was about to make love to his wife.

IT'S RAINING BABIES

In Detroit in the 1930s, a man named Joseph Figlock was walking down the street when he was struck by a baby that had fallen out of a high window. The baby's fall was broken and both man and baby were unharmed. A year later, the same baby fell from the same window onto the unsuspecting Figlock as he was again passing beneath. Again both survived the experience.

POKER FACE

When Robert Fallon was shot dead at the table of San Francisco's Bella Union Saloon in 1858 for cheating at poker, a call was put out on the street for a man to take his place. It was considered unlucky to take winnings—$600 in this case—acquired by cheating. A young man took the dead player's chair, but far from being the pushover he looked, he turned the $600 into $2,200. When the police asked him for $600 to give to the dead man's next of kin, the young man coolly replied that he was the next of kin. He hadn't seen his father for seven years and just happened to be walking past the bar at the time.

And finally this much recorded story. Wouldn't it be neat if it were true? Though not, of course, for Henry Ziegland:

THE WORLD'S SLOWEST BULLET

A bullet fired at Henry Ziegland didn't strike and kill him until twenty years later.

In 1893, Ziegland broke off a relationship with a girlfriend who, distraught, committed suicide. In revenge, the girl's brother tracked down Ziegland and fired at him in the garden of his home. The brother, believing he had killed Ziegland, turned his gun on himself and took his own life. But Ziegland survived. The bullet had merely grazed his face and embedded itself in a tree. In 1913 Ziegland decided to use dynamite to uproot and remove the tree, which still had the bullet embedded in it. The explosion propelled the bullet into Ziegland's head—killing him outright.

THE ULTIMATE
COINCIDENCE

Perhaps we should call it the First Coincidence. Or the Last Coincidence. Either would suit, but "ultimate" most fits its superlative significance. It's the most important coincidence in our life, in everybody's life; in the life of our planet, our solar system and our universe. To begin with, it brought us all together. It's the reason we are. And if ever its felicitous consonance should alter, we won't be around to speculate whether it was a happy accident or part of a grand unified design. Nothing will be around.

We're talking about fundamentals here; the fundamental physical laws pertaining to the day-to-day running of the universe. Physicists call them the fundamental constants—things like the masses of atomic particles, the speed of light, the electric charges of electrons, the strength of gravitational force.... They're beginning to realize just how finely balanced they are. One flip of a decimal point either way and things would start to go seriously wrong. Matter wouldn't form, stars wouldn't twinkle, the universe as we know it wouldn't exist and, if we insist on taking the selfish point of view in the face of such spectacular, epic, almighty destruction, nor would we.

The cosmic harmony that made life possible exists at the mercy of what appear, on the face of it, to be unlikely odds.

Who or what decided at the time of the Big Bang that the number of particles created would be 1 in 1 billion more than the number of antiparticles, thus rescuing us by the width of a whisker from annihilation long before we even existed (because when matter and antimatter meet, they cancel each other out)? Who or what decided that the number of matter particles left behind after this oversize game of cosmic swapping would be exactly the right number to create a gravitational force that balanced the force of expansion and didn't collapse the universe like a popped balloon? Who decided that the mass of the neutron should be just enough to make the formation of atoms possible? That the nuclear force that holds atomic nuclei together, in the face of their natural electromagnetic desire to repulse each other, should be just strong enough to achieve this, thus enabling the universe to move beyond a state of almost pure hydrogen?

Who made the charge on the proton exactly right for the stars to turn into supernovas? Who fine-tuned the nuclear resonance level for carbon to just delicate enough a degree that it could form, making life, all of which is built on a framework of carbon, possible?

The list goes on. And on. And as it goes on—as each particularly arrayed and significantly defined property, against all the odds, and in spite of billions of alternative possibilities, combines exquisitely, in the right time sequence, at the right speed, weight, mass, and ratio, and with every mathematical quality precisely equivalent to a stable universe in which life can exist at all—it adds incrementally in the human mind to a growing sense, depending on which of two antithetical philosophies it chooses to follow, of either supreme and buoyant confidence, or humble terror.

The first philosophy says this perfect pattern shows that the universe is not random; that it is designed and tuned, from the atom up, by some supreme intelligence, *especially* for the purpose of supporting life.

The other says it's a one in a trillion coincidence.

INDEX